SpringerBriefs in Environment, Security, Development and Peace

Volume 1

Series Editor

Hans Günter Brauch

For further volumes:
http://www.springer.com/series/10357

Mely Caballero-Anthony
Youngho Chang · Nur Azha Putra
Editors

Energy and Non-Traditional Security (NTS) in Asia

Springer

Editors
Mely Caballero-Anthony
Centre for Non-Traditional Security (NTS) Studies
S. Rajaratnam School of International Studies
Nanyang Technological University
Singapore 639798
Singapore

Nur Azha Putra
Energy Studies Institute
Energy Security Division
National University of Singapore
Block A, #10-01
29 Heng Mui Keng Terrace
Singapore 119620
Singapore

Youngho Chang
Division of Economics
Nanyang Technological University
14 Nanyang Drive
Singapore 637332
Singapore

ISSN 2193-3162 ISSN 2193-3170 (electronic)
ISBN 978-3-642-29705-2 ISBN 978-3-642-29706-9 (eBook)
DOI 10.1007/978-3-642-29706-9
Springer Heidelberg New York Dordrecht London

Library of Congress Control Number: 2012938951

Printed on acid-free paper

Springer is part of Springer Science+Business Media (www.springer.com)

Preface

This volume is a compilation of selected papers presented at a conference on *Energy and Non-Traditional Security*. The conference, held in August 2008 in Singapore, was organised by the Centre for Non-Traditional Security (NTS) Studies at the S. Rajaratnam School of International Studies, Nanyang Technological University, Singapore.

The volume was first conceptualised against the backdrop of the world oil crisis of 2007–2009 and the impacts of the steep price rises on human collectivities. Since the convening of the conference, many significant global events have occurred, reinforcing further the salience of understanding the implications of NTS. The Arab Spring revolutions and the Fukushima nuclear disaster of 2011, for instance, have only increased awareness among the security community and policymakers of the scale of NTS challenges, as well as the multiplicity and cross-cutting implications of the threats, and the pressing need to address them.

Mely Caballero-Anthony and Nur Azha Putra introduce the volume with "Energy and Non-Traditional Security (NTS)—Understanding Security from Below". This chapter introduces the concept of NTS, and provides the conceptual framework which acts as the theoretical foundation for this volume and its companion, *Rethinking Energy Security: A Non-Traditional View of Human Security*. The NTS framework seeks to expand the notion of energy security, from one anchored to a traditional, state-centric approach, to one based on human security or the security of human collectivities. This chapter unpacks the basic principles of energy security and reconstructs it through the NTS framework. In doing so, the authors open up the space for analyses of energy policies that are informed by various NTS themes and perspectives.

Youngho Chang and Swee Lean Collin Koh in "Rethinking Market Governance and Energy Security" discuss how the market, reinforced with good governance, regulate and normalise the impact of rising global oil prices on the public. For instance, they argue that good market governance, which promotes transparency, accountability and respect for competitive policies, would attract potential investors to fund long-term solutions (such as alternative energy technologies) to energy security problems. Hence, in this chapter, Chang and Koh stretch the notion of energy security to include market governance issues.

In "Energy Security and Climate Change in ASEAN: Implications and Policies", Youngho Chang and Lixia Yao present a quantifiable definition of energy security and illustrate how to measure energy security. The authors apply the quantifiable definition to Association of Southeast Asian Nations (ASEAN) countries, evaluating the policies and activities of these countries as they attempt to ensure energy security. Key policies and activities discussed are infrastructure building and hydropower exploitation in the region. They also discuss current efforts to enhance energy security and mitigate climate change, noting that the latter has profound impacts on the former. The authors also suggest how to address issues related to energy security and climate change in the region.

Maria Nimfa F. Mendoza in "The Socioeconomic Impact of Energy Security in Southeast Asia" examines the impact of rising energy prices in the global market on the various economies and communities in Southeast Asia. She also surveys the energy policies of Southeast Asian states and their attempts to mitigate rising fuel and energy prices, and how various communities have adapted to the higher costs of living. She concludes by arguing that the general tendency among states to aim for energy self-sufficiency and independence (in response to the spike in the global oil prices) may not, in the long run, actually lead to energy security for states in the region. Instead, she observes that energy security for Southeast Asian states may be best achieved through regional cooperation and interdependence.

In "An Environmental Perspective on Energy Development in Indonesia", Fitrian Ardiansyah, Neil Gunningham and Peter Drahos discuss the impact of Indonesia's energy security policy on its environment and economic activities, with particular reference to its shift towards coal and biofuels. For instance, they argue that the development of biofuels, particularly from palm oil, would lead to the rapid clearing of tropical forests, and this would affect forest-dependent people and wildlife. Therefore, they call for a more sustainable approach to the development of biofuels, suggesting, for instance, the harvesting of non-forested and abandoned lands, which can actually mitigate climate change and enhance economic development.

This volume on *Energy and Non-Traditional Security in Asia*, together with its accompanying volume, *Rethinking Energy Security: A Non-Traditional View of Human Security*, aims to widen the debate on energy security beyond the conventional views of what energy security means to the security and well-being of states and societies. We hope that the two volumes would encourage a more robust debate and lead to a more comprehensive approach to dealing with energy security challenges. We also hope that in advancing an NTS perspective to this complex security challenge, we can meaningfully contribute to the promotion of human security in Asia and beyond.

Mely Caballero-Anthony

NTS	Non-traditional security
ASEAN	Association of Southeast Asian Nations

Editorial

Springer Briefs in Environment, Security, Development and Peace (ESDP)

Hans Günter Brauch[1]

This book series was stimulated by a chapter in the *Hexagon Series on Human and Environmental Security and Peace* (HESP) on the "Conceptual Quartet: Security and its Linkages with Peace, Development and the Environment".[2] This addressed the conceptual history and the context of four key concepts at the centre of four major research programmes in the social and political sciences and in international relations: (a) environmental studies or ecology; (b) international, national and human security studies; (c) development studies; and (d) peace studies or peace research. Six chapters analysed the dyadic conceptual relationship between: (i) peace and security; (ii) peace and environment; (iii) development and security; (iv) environment and development (sustainable development); (v) development and security; and (vi) environment and security.

These four key concepts and the six dyadic conceptual relationships address significant issues in international politics and in international relations but also in many other disciplines in the social and natural sciences. Both the HESP series of major multidisciplinary handbooks and these *SpringerBriefs in Environment, Security, Development and Peace* (ESDP) address the linkages across disciplines.

---

[1] Hans Günter Brauch (Dr. phil., Heidelberg University; Dr. habil., Free University of Berlin, Germany) has taught as a Privatdozent (Adj. Prof.) at the Otto-Suhr Institute for Political Science of the Free University of Berlin since 1999; since 1987 he has been Chairman of Peace Research European Security Studies (AFES-PRESS), an independent non-governmental and tax-exempt international scientific society in Mosbach in South Germany. He also occasionally teaches at SciencePo in Paris, at the European Peace University (EPU) in Schlaining (Austria), and the PhD Programme of the *Centro de Estudios Superiores Navales* (CESNAV) in Mexico. Since 2006 he has been a senior fellow with the *United Nations University's Institute on the Environment and Human Security* (UNU-EHS).

[2] Hans Günter Brauch, 2008: "Conceptual Quartet: Security and its Linkages with Peace, Development and Environment", in: Brauch, Hans Günter; Oswald Spring, Úrsula; Mesjasz, Czeslaw; Grin, John; Dunay, Pal; Behera, Navnita Chadha; Chourou, Béchir; Kameri-Mbote, Patricia; Liotta, P.H. (Eds.): *Globalization and Environmental Challenges: Reconceptualizing Security in the twenty-first Century*. Hexagon Series on Human and Environmental Security and Peace, vol. 3 (Berlin–Heidelberg–New York: Springer): 65–98.

In both the HESP and the ESDP series, the editor welcomes book proposals and completed manuscripts by natural and social scientists, as well as by multidisciplinary teams of authors. The material may address issues of global change and its impacts on humankind, on environmentally induced migration, on crises and conflicts, as well as proposing cooperative strategies to cope with these challenges either locally or in the framework of international organizations and regimes.

While the HESP series offers a platform for scientific communities dealing with global environmental and climate change, disaster reduction, human, environmental and gender security, and peace and conflict research, as well as for humanitarian aid and the policy community in national governments and international organizations, the focus of *SpringerBriefs in Environment, Security, Development and Peace* (ESDP) series is wider.

These *SpringerBriefs* present concise summaries of cutting-edge research and innovative policy perspectives. The series focuses on the interconnection of the new and non-traditional global environmental and development challenges that face humankind and that may pose dangers to peace and security in the Anthropocene era of the Earth's history. *SpringerBriefs in ESDP* publish short monographs and edited volumes of workshops that are peer reviewed by scholars from many disciplines and from all parts of the world. These *SpringerBriefs* will give more 'voice' and 'visibility' to scientists and innovative political thinkers in developing countries. *SpringerBriefs in ESDP* provide multi-, inter- and transdisciplinary knowledge relevant for coping with the numerous projected impacts of global environmental change, political, economic and cultural crises and conflicts. They offer

- Timely state-of-the art analyses (e.g. of earth systems science, geoecology) and policy assessments of the global challenges facing humankind in the twenty-first century.
- Bridges between new research results with snapshots of hot and/or emerging topics, literature reviews and in-depth case studies.

As the *SpringerBriefs in ESDP* evolve, several subseries are planned on the four issue areas of environment, security, development and peace. The first will be the journal *Mediterranean Soil Ecosystems* that focuses on interdisciplinary research dealing with land issues of soil, water and environmental management. This journal is edited by Prof. Dr. Selim Kapur (University of Adana, Turkey) with the support of an editorial board that will manage the anonymous peer-review process.

The series editor welcomes brief concept outlines and original manuscripts as proposals. If they are considered of relevance, these proposals will be peer reviewed by experts from the natural and social sciences and from the humanities.

Mosbach, Germany, 21 April 2012 Hans Günter Brauch

Editor, SpringerBriefs in Environment, Security, Development and Peace (ESDP)
Editor, SpringerBriefs on Pioneers in Science and Practice (PSP)
Editor, Hexagon Series on Human and Environmental Security and Peace (HESP)

Contents

Contents

Chapter 1
Introduction: Energy and Non-Traditional Security (NTS)—Understanding Security from Below

Mely Caballero-Anthony and Nur Azha Putra

Abstract The pursuit of energy security has long been associated with intense rivalry and competition among states for energy resources that are available in limited amounts outside their borders. More often than not, this competition has been perceived as a zero-sum game where one state's security can only be achieved at the expense of another's. Since the end of the Cold War and with it, the decline of inter-state wars, the idea that security primarily revolves around the sovereignty of the state has been challenged by the emergence of non-traditional security (NTS) scholars who argue that the debate should shift towards human collectivities instead.

Keywords Security · Non-traditional security · Energy security · Energy efficiency · Sustainable development · Renewables · Nuclear energy

---

M. Caballero-Anthony (✉)
Centre for Non-Traditional Security (NTS) Studies,
S. Rajaratnam School of International Studies (RSIS),
Nanyang Technological University (NTU), Block S4, Level B4,
Nanyang Avenue, Singapore 639798, Singapore
e-mail: ISMCAnthony@ntu.edu.sg
URL: http://www.rsis.edu.sg/about_rsis/staff_profiles/Mely_Anthony.html;
http://www.rsis.edu.sg/nts

N. Azha Putra
Energy Studies Institute, 29 Heng Mui Keng Terrace, Block A, #10-01,
Singapore 119620, Singapore
e-mail: azha@nus.edu.sg
URL: http://www.esi.nus.edu.sg/our-researchers/nur-azha-putra

M. Caballero-Anthony et al. (eds.), *Energy and Non-Traditional Security (NTS) in Asia*, SpringerBriefs in Environment Security, Development and Peace 1,
DOI: 10.1007/978-3-642-29706-9_1,

1.1 Introduction

Traditional notions of security are founded on the belief that national security concerns premised on military threats to state security supersede other security threats that could pose grave threats to the state and human security. The traditional security perspective puts the protection of a state's sovereignty from the threat of wars and violence as its sole concern. However, this perspective has been challenged by a number of scholars who have sought to broaden the security discourse by challenging the dominant belief that security revolves mainly around military threats.

Among the security analysts that have challenged the traditional security perspective are those that advanced the field of non-traditional security (NTS). NTS scholars argue that since the end of the Cold War and the collapse of the Soviet Union, threats to national security have been increasingly non-military in nature, with more incidents of internal and cross-border conflicts than inter-state wars occurring on the world stage. For instance, Caballero-Anthony et al. (2006) observe that security threats have in recent times evolved from military confrontations to non-military conflicts such as food security, water security, climate change, human trafficking, pandemics, cross-border conflicts and energy security. They add that most of these challenges are transnational in terms of their origins, conception and effects. Caballero-Anthony et al. (2006) argue that, more importantly, NTS has shifted the focus of security beyond the state to include human collectivities as the primary referent object of security.

Explaining the dynamics between traditional and non-traditional security, Hough (2008) points out that traditional realists regard non-military issues as "low politics". To these scholars, military, and not non-military, issues are what governments must prioritise and accord "national security" status. This perspective was eventually contested as being too narrow. Clearly responding to the observation that states are vulnerable also to non-military threats, an idea that resonates with the United Nations' notion of human security, several security scholars made the case that non-military issues should also be securitised and framed as a matter of national security (Buzan et al. 1998; Caballero-Anthony et al. 2006; Tow et al. 2000). According to Hough (2008), widening the concept of security is not simply a move to reconceptualise the meaning of national interests and reframe "high politics". Rather, it is an attempt to expand the security dimension to include non-military threats to human collectivities.

Therefore, "whose security are we referring to?" becomes the primary question. The answer can no longer be presumed to be the security of the state. Most recently, there is sufficient evidence, drawing on the cases of Tunisia's Jasmine Revolution and the Arab Spring revolutions in the Middle Eastern and North African (MENA) countries in 2011, to reasonably conclude that states can be brought down even in the absence of war and inter-state conflicts. A common thread running through the street protests and demonstrations is human security issues such as social and political freedom, poverty and economic development.

This volume is a valuable addition to the literature on energy security as it aims to examine the implications of energy security from a non-traditional perspective. The chapters and their analyses of energy security go beyond the traditional preoccupation with the security of supply, and concerns such as the security of infrastructure, maritime security, and geopolitical tensions between energy producing, consuming and transit countries. Instead, the volume emphasises the discussion of energy security through a human security lens in order to decipher their implications for the well-being of states and societies. Thus, the themes that are covered in this collection highlight issues such as governance and socioeconomic impacts. The chapters also draw the linkages between energy security, and climate change, health, environmental degradation and energy sustainability. Moreover, in analysing the implications of energy security for inter-state relations, this volume underscores the importance of promoting cooperation instead of competition, and of fostering regional cooperation.

The following sections provide a more detailed discussion of the salient elements of energy security from an NTS perspective.

1.2 NTS Perspectives on Energy Security

Essentially, energy security holds different significance for different nations, with the meanings largely dependent on whether the countries are exporters or importers of energy resources, and on the level of their economic development. Therefore, the notion of energy security varies by nation and region, and by their economic status.

According to the International Energy Agency (IEA), energy security refers to "uninterrupted physical availability at a price which is affordable, while respecting environment concerns".[1] Yergin articulates a definition with an international relations nuance: energy security is "lodged in the larger relations among nations and how they interact with one another".[2] Others such as Brown et al. (2003, p. 7) define energy security in terms of a resilient energy system which is able to withstand security threats.

Notwithstanding the different definitions and contexts, it could be suggested that, in general, energy security is about access to an adequate, affordable and sustainable supply of energy to meet the needs and aspirations of private and public consumption, and the provision of that supply in a way that is environment-friendly and takes into account the continued availability of energy resources for the consumption of future generations. NTS perspectives build upon this

[1] See IEA (International Energy Agency), 2011, "Energy Security", at: http://www.iea.org/subjectqueries/keyresult.asp?KEYWORD_ID=4103 (1 December 2011).

[2] Yergin, Daniel, "Ensuring Energy Security", in: *Foreign Affairs* (March/April 2006), at: http://www.foreignaffairs.com/articles/61510/daniel-yergin/ensuring-energy-security (1 December 2011).

conception and add that energy security should also enhance the security of human collectivities at all levels. To that end, NTS scholars call for collaboration rather than competition among states. The NTS approach therefore holds that the discussion of energy security must span three interdependent levels: international, national and individual. At the international level, energy security encompasses issues among nations. At the national level, it is about the ability of the state to supply energy affordably, reliably and sustainably. At the individual level, energy security is about the ability to maximise the economic utility of energy sources. Using an NTS framework, there are three elements to achieving energy security: security, stability and sustainability.

1.2.1 Security

From a geopolitics perspective, the energy security policies adopted by countries will have a direct bearing on the relations among nations. NTS scholars argue that the emphasis of such policies has to shift from competition to collaboration. Rather than competing for scarce resources such as oil and gas, they call for greater collaboration among nations within international and regional frameworks.

Nations are also encouraged to work together to manage the security of transnational energy infrastructures. Gas pipelines, and the transportation and storage of oil and gas, typically run across state borders and along international sea routes. The security of these transnational facilities and maritime routes requires collaboration among state security agencies. Inevitably, such frameworks would require greater dialogue, cooperation and information sharing among the stakeholders in the countries involved. Perhaps of greater significance is the potential for these frameworks to have a positive impact on regional stability and international security. It is thus important that a regional or international energy-security paradigm exists as a platform to facilitate greater cooperation and collaboration among the stakeholders.

1.2.2 Stability

Stability in the supply and demand of energy is particularly critical in ensuring that nations are protected from any supply shortages and rise in demand, and the concomitant price increases. Prolonged energy price hikes could lead to social unrest. For instance, in 2008, hundreds of villages in the Central Asian state of Tajikistan faced dire conditions when the price of electricity was raised by 20%, to a level barely affordable to the villagers.[3] The situation worsened during winter

[3] Najibullah, Farangis, "Tajikistan: Energy Shortages, Extreme Cold Create Crisis Situation", in: *Eurasianet.org* (12 January 2008), at: http://www.eurasianet.org/departments/insight/articles/pp011308.shtml (14 June 2009).

when the country had to rely on imported electricity from its neighbouring countries, Kyrgyzstan, Uzbekistan and Turkmenistan who themselves frequently run into energy shortages. Tajikistan's hydroelectric dams could not produce enough energy due to insufficient generating capacity and outdated facilities.[4]

In Europe, the soaring fuel prices of 2007–2008 brought waves of street protests and demonstrations in France, Spain, Portugal and Italy. Workers in industries such as fishing, trucking and agriculture were badly hit by the oil crisis. They went on strike and demanded that their governments reduce the price of fuel. Fishermen in Spain, Portugal and Italy, for instance, demanded that their governments provide subsidies to address the gap between high oil prices and low prices for fish.[5]

In Asia, the Indonesian government removed its subsidies for fuel, which led to a drastic price hike of almost 30 % in May 2008. This move was met by a series of street protests.[6] Interestingly, this happened a few years after Indonesia became a net importer of crude oil in 2005. Around the same time, in 2008, the Malaysian government, which also provided fuel subsidies, took a different path and imposed limits on fuel sales to Singapore-registered cars. The government felt that its subsidies should not be extended to non-Malaysians.[7] Meanwhile, several airlines in the United States were forced to reduce the number of flights and had to retrench staff as a way of lowering their operational costs, among them Continental Airlines, American Airlines, Delta Air Lines and United Airlines.[8] In South Korea, the Transportation Industry Nationwide Labour Union of Korea went on strike and demanded that the government lower diesel prices and increase transportation fees.[9]

1.2.3 Sustainability

Sustainability is another key concept in examining NTS threats. The world's reliance on traditional fossil fuels such as oil, gas and coal has led to fears that those fuels would eventually be depleted. The high consumption of fossil fuels also raises

---

[4] Marat, Erica, "Tajikistan Experiences Rolling Blackouts amid Freezing Winter Temperatures", in: *Eurasia Daily Monitor*, 5,14 (24 January 2008), at: http://www.jamestown.org/single/?no_cache=1&tx_ttnews%5Btt_news%5D=33322 (14 June 2009).

[5] "Europe Fuel Protests Spread Wider", in: *BBC News* (30 May 2008), at: http://news.bbc.co.uk/2/hi/7426971.stm (1 December 2011).

[6] Deutsch, Anthony; Hidayat, Taufan, "Indonesian Parliament Backs Fuel Subsidy Cut", in: *Financial Times* (14 December 2010), at: http://www.ft.com/cms/s/0/f37cb472-078e-11e0-8d80-00144feabdc0.html#axzz1gyHf5yrT (18 December 2011).

[7] "Malaysia to Ban Border Fuel Sales to Foreigners", in: *channelnewsasia.com* (27 May 2008), at: http://www.channelnewsasia.com/stories/afp_asiapacific/view/350252/1/.html (14 May 2009).

[8] "Continental Airlines to Cut 3,000 Jobs, Reduce Flights", in: *St. Louis Business Journal* (5 June 2008), at: http://www.bizjournals.com/stlouis/stories/2008/06/02/daily47.html (14 December 2011).

[9] Myung, Bryan, "Transport Strike in Yeosu Affects Plants", in: *ICIS.com* (12 June 2008), at: http://www.icis.com/Articles/2008/06/12/9131791/transport-strike-in-yeosu-affects-plants.html (18 December 2011).

questions of environmental sustainability, particularly as high carbon emissions have been implicated in climate change. To address the potential depletion of traditional energy sources and to mitigate the impact on the environment, countries have turned to various strategies, such as promoting energy efficiency, and pursuing alternative energy sources such as renewables and nuclear energy.

1.2.3.1 Energy Efficiency: The Fifth Fuel

Energy efficiency has, quite rightly, been referred to as the fifth energy fuel. The other four are coal, natural gas, nuclear energy and renewables. The American Council for an Energy-Efficient Economy defines energy efficiency as "a means of using less energy to provide the same (or greater) level of energy services" (Furrey and Black 2009, p. 3). The sectors consuming the most amount of energy are typically buildings, industries, utilities and transportation. Naturally, the call for greater energy efficiency would affect these sectors the most.

There are several ways in which greater energy efficiency can be achieved. One is through encouraging energy conservation. This would require a change in the behaviour and attitude of users of electricity, for instance, they would have to switch off the lights when they leave the office. The second is through the use of products incorporating green technology. This approach is now increasingly employed by the private sector. Measures adopted include using energy-saving office equipment such as air-conditioners, printers and computers. There is still room for improvement in this area. For instance, a 2011 survey by Singapore's National Environment Agency found that a fifth of air-conditioner models and 9 per cent of refrigerators do not meet the minimum energy-efficiency standards specified under the Environmental Protection and Management Act.[10] In the transportation sector, motorists could be encouraged to opt for electric or hybrid motor vehicles, which use cleaner forms of energy and are thus more environment-friendly.

Singapore is a case in point. It has institutionalised a number of legal frameworks and strategies to encourage industry, homes and the transportation sector to become more energy efficient. In 2010, the Energy Market Authority and the Land Transport Authority initiated the test-bedding of an electric vehicle programme in Singapore.[11] The objective is to examine the infrastructure requirements, and to identify business models arising from the use of electric vehicles.[12] Additionally,

---

[10] Ng, Lian Cheong; Soh, Alvina, "20 % of Air-cons, 9 % of Fridges Don't Meet Energy Standards: NEA", in: *channelnewsasia.com* (31 August 2011), at: http://www.channelnewsasia.com/stories/singaporelocalnews/view/1150310/1/.html (15 December 2011).

[11] Singapore Economic Development Board, "Test-bedding of Electric Vehicles in Singapore from 2010", Press Release (6 May 2009), at: http://www.sedb.com/etc./medialib/downloads/media_release_2009/electric_vehicles.Par.87936.File.tmp/Press%20Release.pdf (21 July 2011).

[12] Ibid.

the Building and Construction Authority has implemented the Green Mark Scheme to drive the country's construction industry towards more environment-friendly buildings.[13]

1.2.3.2 Renewables

Renewables such as solar energy, hydroelectric power, biofuels and wind energy could to some extent be alternatives to fossil fuels. However, they still require extensive investment in research and development before they can be fully commercialised. While renewables have enormous potential, the returns on investment are somewhat uncertain due to a number of economic factors.

Nevertheless, three countries in Southeast Asia look set to pursue at least one renewable-energy source—hydropower. Dore and Yu (2004, p. 4) estimate that the upper Mekong River has the potential to produce about 28,930 MW while the Mekong River Commission (2010) estimates that the lower Mekong River has the potential to produce at least 30,000 MW of electricity. It makes sense then that Laos, Cambodia and Myanmar have announced plans to develop and harness that hydropower potential. Situated along the lower Mekong River, these countries intend, with financing from the Asian Development Bank and other foreign investors, to construct dams and pico-hydropower projects. In particular, Laos aims to be the battery of Southeast Asia (Bounthongvongsaly et al. 2010). The government plans to produce hydroelectricity for commercial purposes; it intends to sell its hydroelectricity to its energy-hungry neighbours, Thailand and Vietnam. Electricity from its numerous pico-hydropower projects will be used to provide power for a large part of its rural population. Some commentators estimate that there could already be as many as 200 dams which are either completed, currently under construction or planned along the Mekong River.[14]

1.2.3.3 Nuclear Energy

Nuclear energy has been successfully adopted in many countries over the last five decades, especially in North America, Western Europe, Japan, Korea and China. It thus appears to Southeast Asian states to be a viable option. As such, several countries in the region have, in recent times, expressed interest in developing their own nuclear energy facilities, or have announced plans to do so.

The nuclear disaster in Fukushima, Japan, in March 2011, however, demonstrated the vulnerability of nuclear power reactors to NTS threats. The Fukushima

[13] Building and Construction Authority of Singapore, 2011, "About BCA Green Mark Scheme", at: http://www.bca.gov.sg/greenmark/green_mark_buildings.html (21 July 2011).

[14] "Hydropower Development and Mekong River Fisheries: What Can Be Learned from the Columbia River?", in: *International Water Power & Dam Construction* (5 October 2011), at: http://www.waterpowermagazine.com/story.asp?storyCode=2060836 (12 October 2011).

Daiichi nuclear power plant's infrastructure and boiling water reactors were designed to withstand earthquakes measuring up to 8.2 on the Richter scale and a 20-foot tsunami; the resulting 9.0-magnitude earthquake and over 30-foot tsunami proved to be too much for the plant's security features (Yeo 2011). Japan's Nuclear and Industrial Safety Agency classified the disaster at level 7 on the International Nuclear and Radiological Event Scale. This puts it on par with the Chernobyl incident in 1986. The 1979 Three Mile Island incident in Pennsylvania, United States, was categorised level 5. Consequently, electricity supply to the north-eastern and eastern regions of Japan was disrupted. Thousands of households were left with rolling blackouts and minimal heat. The transportation network was also severely hit as a result of train service disruptions.

The reactions of Southeast Asian governments were predictable. They were quick to revise their energy security policies and their plans to adopt nuclear energy. President Susilo Bambang Yudhoyono of Indonesia said, "What happened in Japan last March can happen in Indonesia because (the two countries') geography is very much similar."[15] On 27 April 2011, Thailand's Energy Minister, Wannarat Channukul, said that the country has delayed its first nuclear power plant project by 3 years due to public concerns. It had previously planned to start a commercial nuclear power plant in 2020.[16] As for the Philippines, it is unlikely that the government will pursue any nuclear options in the short- to medium-term due to strong domestic opposition. This is despite indications in the Philippine Energy Plan for 2007–2014 that there are plans to revive the Bataan Nuclear Power Plant as well as construct four more nuclear plants between 2015 and 2025. These nuclear power plants had been expected to provide up to about 3,020 MW in generation capacity for the Philippines by 2035 (IAEA 2011, Table 8).

Only one country remained undeterred. While other countries in Southeast Asia rethought their nuclear power plans in the wake of Fukushima, Vietnam stood behind its ambitious plans. Vietnam intends to build at least eight to fourteen nuclear facilities in the next two decades.[17]

In the larger scheme of things, the euphoria and apparent rush among Southeast Asian states to adopt nuclear energy, which began sometime in the middle of the last decade, appear to have waned as a consequence of the Fukushima nuclear disaster. This could perhaps have been due to the negative public perception of the safety of nuclear plants, strong lobbying by green activists and the inability of states to allay public concerns about their nuclear ambitions. Another possibility

---

[15] "Indonesian President Backs Away from Nuclear Power Plant Proposal", in: *Jakarta Globe* (18 June 2011), at: http://www.thejakartaglobe.com/business/indonesian-president-backs-away-from-nuclear-power-plant-proposal/447676 (19 June 2011).

[16] Jittapong, Khettiya, "Thailand Delays First Nuclear Power Plant to 2023", *Reuters* (27 April 2011), at: http://uk.reuters.com/article/2011/04/27/thailand-nuclear-idUKL3E7FR2WB20110427 (30 January 2012).

[17] "Japan Firm to Study Vietnam Nuclear Plant", *AsiaOne* (28 September 2011), at: http://www.asiaone.com/News/Latest+News/Asia/Story/A1Story20110928-302045.html (29 September 2011).

that is worth pondering is the influence of the Arab Spring revolutions on these decisions. The Fukushima disaster occurred at a time when the Arab Spring revolutions were gaining traction in the MENA region. There was widespread coverage in the mainstream media and on social networking forums. Presumably, states in Southeast Asia found it necessary to appeal to popular public sentiments and thus adopted a cautious stance on the use of nuclear energy. Interestingly, none declared outright that they would abandon their nuclear energy ambitions.

In June 2011, Singapore's Foreign Affairs Minister, K. Shanmugam, speaking at the Asia-Europe Meeting (ASEM) in Hungary, proposed that Singapore host an ASEM seminar on nuclear safety issues in 2012. He felt that the Fukushima disaster had opened up the path for countries aspiring to develop nuclear energy to share expertise and experience on safety issues.[18] Presumably, this suggestion was well-received by Southeast Asian states since in September 2011, Singapore's Ministry of Foreign Affairs confirmed that it will host a regional seminar in 2012 to discuss cooperation, emergency preparedness and response capabilities.[19]

Singapore's attempts following the Fukushima disaster to galvanise the region towards a common shared objective in the area of nuclear energy security seem to underscore Caballero-Anthony et al.'s (2006) observations that security threats have in recent times evolved from military confrontations to non-military conflicts such as climate change and energy security, and that the non-military nature of such NTS threats would lead to states being more open about collaboration and cooperation in the area of shared security.

1.2.4 Conclusion

In conclusion, the traditionalists' notion of energy security has come under intense scrutiny and has been broadened by NTS scholars to include other referent objects beyond the state. Through the lens of human security which focuses on the general welfare and well-being of human collectivities, proponents of NTS argue that energy security is not simply about the security and sovereignty of the state, it is also about ensuring the well-being and dignity of its people as they respond to and manage the consequences of energy scarcity. A human-centred approach to energy security allows for a more sustainable approach to meeting energy challenges now and in the future, one which takes into consideration the needs of other peoples across a state's borders. Hence, sustainable development is an important feature of

---

[18] Chan, Sabrina, "Shanmugam Proposes Singapore Host Nuclear Safety Seminar", *channelnewsasia.com* (7 June 2011), at: http://www.channelnewsasia.com/stories/singaporelocalnews/view/1133779/1/.html (9 August 2011).

[19] Ministry of Foreign Affairs Singapore, "Minister for Foreign Affairs and Law K Shanmugam at the 66th Session of the United Nations General Assembly, 22 September 2011", Press Release (23 September 2011), at: http://app.mfa.gov.sg/2006/press/view_press.asp?post_id=7272 (2 October 2011).

the NTS approach to energy security. To achieve this requires no less than greater cooperation and collaboration between states and among their peoples.

References

Bounthongvongsaly T, Susanto J, Smits M, Nanthavong K, Gaillard L (2010) Policy brief on pico hydropower. Lao Institute for Renewable Energy, Vientiane

Brown MH, Rewey C, Gagliano T (2003) Energy security. National Conference of State Legislatures, Denver

Buzan B, Waever O, de Wilde J (1998) Security: a new framework for analysis. Lynne Rienner Publishers, Boulder

Caballero-Anthony M, Emmers R, Acharya A (eds) (2006) Non-traditional security in Asia: dilemmas in securitisation. Ashgate, London

Dore J, Yu X (2004) Yunnan hydropower expansion: update on China's energy industry reforms and the Nu, Lancang and Jinsha hydropower dams, Working Paper, Chiang Mai University's Unit for Social and Environmental Research and Green Watershed, Chiang Mai and Yunnan

Furrey L, Black S (2009) Energy efficiency resource standards: a state model. American Council for an Energy-Efficient Economy, Washington, DC

Hough P (2008) Understanding global security, 2nd edn. Routledge, Abingdon

IAEA (International Atomic Energy Agency) (2011) Philippines (Profile added in August 2009). Country nuclear power profiles—2011 Edition. http://www-pub.iaea.org/mtcd/publications/pdf/cnpp2009/countryprofiles/Philippines/Philippines2009.htm. 28 April 2011

Mekong River Commission (MRC) (2010) State of the basin report 2010. MRC, Vientiane

Tow WT, Thakur R, Hyun I-T (eds) (2000) Asia's emerging regional order: reconciling traditional and human security. United Nations University Press, Tokyo and New York

Yeo D (2011) The Fukushima Daiichi nuclear reactor crisis: what happened. ESI Bull 4(1):(Singapore: Energy Studies Institute): 2. http://www.esi.nus.edu.sg/news/2011/09/12/fukushima-nuclear-disaster-(volume-4-issue-1)—april-2011. 9 Aug 2011

Abbreviations

ASEM	Asia-Europe Meeting
ESI	Energy Studies Institute
IEA	International Energy Agency
MENA	Middle Eastern and North African
MRC	Mekong River Commission
MW	Megawatt
NTS	Non-traditional security

Author Biographies

Mely Caballero-Anthony (Singapore): Associate professor at the S. Rajaratnam School of International Studies (RSIS), Nanyang Technological University (NTU), Singapore, and head of the RSIS Centre for Non-Traditional Security (NTS) Studies. She has also served as director of external relations in the Political-Security Community Department, Association of Southeast Asian Nations (ASEAN)

Secretariat. Her research interests include regionalism and regional security in the Asia-Pacific, multilateral security cooperation, politics and international relations in ASEAN, conflict prevention and management as well as human security. Her latest publications, both single-authored and co-edited, include: *Political Change, Democratic Transitions and Security in Southeast Asia* (Abingdon: Routledge, 2010); *Understanding Non-Traditional Security in Asia: Dilemmas in Securitization* (UK: Ashgate, 2006); (with Acharya, A.; Emmers, R.): *Studying Non-Traditional Security in Asia: Trends and Issues* (Singapore: Marshall Cavendish, 2006); *Regional Security in Southeast Asia: Beyond the ASEAN Way* (Singapore: ISEAS, 2005); (with Acharya, A.): *UN Peace Operations and Asian Security* (Abingdon: Routledge, 2005).

Nur Azha Putra (Singapore): Research associate at the Energy Security Division of the Energy Studies Institute, National University of Singapore. He was previously associate research fellow at the Centre for Non-Traditional Security (NTS) Studies at the S. Rajaratnam School of International Studies, Nanyang Technological University (NTU), Singapore, where he was the lead researcher for the Energy and Human Security programme and also led research on a range of other NTS issues. He holds an MSc in international political economy from NTU; and a Bachelor of Information Technology from Central Queensland University, Australia.

Chapter 2
Rethinking Market Governance and Energy Security

Youngho Chang and Swee Lean Collin Koh

Abstract Energy security is no longer just a matter of securing access to adequate energy supplies—it is concerned with other aspects as well, such as the environmental and social costs of energy use. This chapter presents the principles of good market governance and argues that proper market governance in the energy sector, combining government regulatory measures and the workings of the free market, would be instrumental in ensuring long-term energy security. Japan is presented as a case country. The chapter illustrates that Japan, by adopting proper market governance in the energy sector, has not only ensured the sustainability of energy supplies but also mitigated accompanying environmental and sociopolitical risks of energy use, albeit the Fukushima accident, which has forced it to review and upgrade its market governance.

Keywords Energy governance · Energy security · Fukushima accident · Principles of market governance

---

Y. Chang (✉)
HSS-04-65, Division of Economics, School of Humanities and Social Sciences,
Nanyang Technological University, 14 Nanyang Drive,
Singapore 637332, Singapore
e-mail: isyhchang@ntu.edu.sg
http://www.hssapps.ntu.edu.sg/faculty/econ.asp?u=isyhchang

S. L. C. Koh
S. Rajaratnam School of International Studies (RSIS),
Nanyang Technological University (NTU),
Block S4, Level B4, Nanyang Avenue,
Singapore 639798, Singapore
e-mail: iscollinkoh@ntu.edu.sg
www.rsis.edu.sg/nts

M. Caballero-Anthony et al. (eds.), *Energy and Non-Traditional Security (NTS) in Asia*, SpringerBriefs in Environment Security, Development and Peace 1,
DOI: 10.1007/978-3-642-29706-9_2, © The Author(s) 2012

2.1 Introduction

Contemporary energy security is concerned not only with the security of supply but also environmental and socioeconomic well-being. Greenhouse gases (GHGs) emitted from the burning of fossil fuels lead to climate change and, in so doing, cause a host of non-traditional security (NTS) problems, such as environmentally induced migration. Still, although energy security is part of international dialogue, such as during the G8 Summit held in Japan in July 2008, the matter of GHG emission cutbacks is yet to be addressed properly.

Unprecedented high energy prices have had widespread effects that have permeated extensively through societies and are posing sociopolitical risks, prompting populist calls for concrete government measures in numerous countries. Civil societies have begun to beseech governments, industries and citizen consumers to pay more attention to environmental and socioeconomic problems associated with energy consumption. However, regardless of such rising calls from certain quarters of the society for a stronger government role, examples of how governance has been overused or abused in the name of energy security do persist. The recent nuclear accident at the Fukushima Daiichi nuclear power plants following the earthquake and tsunami in Japan in March 2011 has caused the decades-long question of the role of governance in ensuring energy security to resurface. The Fukushima disaster shows that tenuous governance between the regulator and those regulated and the lack of a properly functioning energy market can cause unfathomable damage to the economy by thwarting efforts aimed at ensuring energy security.

This chapter argues that proper market governance, combining government regulatory measures and the workings of the free market, would be instrumental in ensuring long-term energy security. Section 2.2 introduces the energy security concept and presents an overview of current global energy trends with respect to the security of supply and the environment. In Sect. 2.3, the pitfalls of market and governance are tabled along with an outline of the principles of good market governance. Section 2.4 discusses the relationship between good market governance and energy security at length. Lastly, Sect. 2.5 provides a concluding summary and puts forth brief policy recommendations.

2.2 Energy Security and Global Trends

2.2.1 The Energy Security Concept

'Energy security' is defined broadly as "a condition in which a nation and all, or most, of its citizens and businesses have access to sufficient energy resources at reasonable prices for the foreseeable future free from serious risk of major disruption of service" (Barton et al. 2004b, p. 5). The concept of energy security is further refined for this chapter as 'an adequate and reliable supply of energy resources at reasonable prices'. The focus of energy security is: (1) limiting vulnerability to disruption given increased dependence on imported energy resources, especially oil from the volatile Middle East region; (2) the provision of adequate supply over time for the increased demand, albeit at reasonable prices; (3) addressing the environmental impact of increased GHG emissions; and (4) the prevention of intentional sabotage to energy infrastructure, particularly in the aftermath of the 9/11 terror bombings in the US (Barton et al. 2004b).

Energy security has critical economic ramifications, nationally and globally. Energy interdependence, as Verrastro and Ladislaw (2007) point out, is the way forward for ensuring energy security, as energy independence is fraught with problems such as the reliability and economic competitiveness of domestic sources vis-à-vis energy imports. A more sophisticated approach to energy policymaking, according to these authors, that fully appreciates global market interdependence and the complex nature of energy security is required to manage policy trade-offs (Verrastro and Ladislaw 2007, p. 95). Energy security is not just confined to the problem of supply and demand but also emerging issues, such as environmental concerns, especially now. This is particularly important taking into account the increasing interconnectedness of countries, implying a need for collective solutions to contemporary energy security problems; as such, ensuring energy security through energy independence may no longer be deemed feasible.

2.2.2 Global Energy Trends: Present and Future

The global energy usage increased by 23 % between 1990 and 2005, with a corresponding 25 % increase in GHG emissions; most of this was on account of developing countries and was largely attributed to the expansion in the transport and service industries (IEA 2008b, p. 15). For the same period, electricity consumption increased globally by 54 % although developing nations continued to rely chiefly on traditional biomass and coal (IEA 2008b). Oil products, according to the International Energy Agency (IEA), remained the most important energy commodity, with a 37 % global share of the total in 2005 (IEA 2008b). According to another IEA study, global oil product demand was expected to reach 87.7 mmbd in 2009, compared to 86.9 mmbd in 2008, representing an annual increase of 1 % (IEA 2008a, p. 4). It remains to be noted, however, that people with undisrupted access to electricity supply are a minority in the world; over a quarter of the world population has no access to electricity even as nearly 2.5 billion people still depend on traditional biomass for their fuel needs (Noe and Pring 2004, pp. 431–434).

The provision of energy ‘have-nots’ with access to modern energy sources, together with a projected increase in global demand for energy resources due to a rapidly growing world population, indicates the need for considerably more energy resources in the future. As Birol (2007) points out, global energy demand is expected to steadily rise at least for the next 25 years. In his reference scenario for the year 2030, Birol (2007, p. 1) projects an annual growth rate of 1.6 % in energy demand, with over 70 % of the total increase being attributed to developing countries.[1] It becomes apparent that this projection is not an exaggeration if it were assumed that socioeconomic development in developing countries would eventually result in increased energy consumption; China and India would be good examples of such a scenario. Fossil fuels are predicted to remain the dominant source of energy until 2030, with demand for coal increasing slightly (Birol 2007).

Regardless of current and projected energy demand patterns, the security of supply is fraught with uncertainties, such as volatile energy prices. While crude oil futures did fall back from the early-July 2008 peak of US$145 per barrel, prices have remained high due to a meagre 2008 stock-build, tight distillate markets and ongoing geopolitical risks (IEA 2008a, pp. 1–53). The last factor, geopolitics, is a major driving force influencing the supply of energy resources. For instance, crude oil prices tipped at US$125 per barrel barely a month after prices hit the ceiling, as tensions in Iran and Nigeria raised supply worries.[2] In the coming years, the world would need to rely on increasingly less accessible, high risk and less reliable fossil fuel sources, such as those in the politically volatile Middle East. Apart from this, contemporary NTS threats, such as terrorism, have added further uncertainties over the safety of energy infrastructure, such as oil refineries and sea lines of communication (SLOC) for oil tankers ferrying the precious commodity from far-flung areas over long distances to consumer nations. The post-9/11 international security environment is certainly not encouraging on the issue. For instance, Alexei Miller, head of Gazprom, a Russian gas giant, recently remarked that the world might face a huge energy shortage after 2012.[3] Though this remains to be seen, and irrespective of the diverse and sometimes-contradictory scientific claims and counterclaims thrown around over energy resource depletion, the security of supply remains a contentious issue for both developed and developing countries and energy producers and consumers alike in a mutually dependent world.

As previously mentioned, the period 1990–2005 saw a rise in not only global energy usage but also GHG emissions. Continual increases in energy demand can therefore be expected to cause increased GHG emissions, leading to global warming. Unsurprisingly, GHG emissions are projected to rise by 55 % between 2004 and 2030 (Birol 2007, pp. 7–9). Scientific and environmental communities have been warning of the dire consequences of climate change, such as rising sea levels from the melting of ice glaciers, while calling for reductions in GHG emissions by urging key emitters to impose concrete measures. Citing a temperature increase of 1.8–4.0 °C in the present century due to global warming, R.K. Pachauri, chairman of the United Nations Intergovernmental Panel on Climate Change (IPCC), beseeched developed countries to cut back on GHG emissions before it was

[1] Demand was, in fact, projected to increase by more than a quarter in 2015 alone.

[2] Tensions were caused by the Iranian nuclear weapons programme and the possibility of a US or Israeli military strike on its facilities as well as due to militant attacks on the Royal Dutch Shell pipelines in the Niger Delta region; see "Oil Creeps above $125 on Iran, Nigeria Tensions", in: *Reuters* (29 July 2008).

[3] "Gazprom Predicts Global Energy Shortage after 2012", in: *Reuters* (23 July 2008).

"too late".[4] This is, however, easier said than done, given the strong differences that persist among various nations. This was reflected in the recently concluded G8 Summit as well, which failed to reach a consensus on numerical emissions targets, but announced a somewhat vague commitment to adopt a 50 % reduction in GHG emissions by 2050.[5] International consensus on GHG emission cutbacks is likely to remain a pipedream in the near future due to a lack of political will. Nonetheless, there are promising prospects for national-level research and development (R&D) initiatives in areas such as alternative energy and energy efficiency technologies for reducing reliance on fossil fuels, and gradual steps are being taken in this direction with regard to long-term energy security too.

2.3 Energy, Market and Governance

This chapter treats governance as the institutional arrangements that govern the transactions of produced goods and services in an economy. There are three institutional factors that are controlled by the bounded rationality and opportunism inherent in the behaviour of economic agents—uncertainty, the specificity of goods and services being traded, and the frequency of transactions. These institutional factors produce four alternative modes of governance—market, bilateral, trilateral and unified. Market governance is an adequate form of governance in which generalised or non-specific goods and services are traded regardless of frequency. Bilateral governance is appropriate for situations where recurring transactions occur at some degree of specificity while trilateral governance is the best form when a third party guarantee is required for securing transactions. Unified governance is required for transactions that occur at very high degree specificity.

2.3.1 Pitfalls of Market and Governance

The end of the Cold War was monumental in attesting to the failure of socialist-style centrally planned economies and the contrasting success of free market economies. Since then, a wave of globalisation has literally swept the world. Still the buzzword today, globalisation has come to affect the way commercial activities

[4] "Cut Carbon Emissions before It's Too Late: Pachauri", in: *The Hindu* (28 July 2008).

[5] The failure to agree on emission targets was due to disagreements between developing nations (which were calling for higher emission cuts by richer developed countries) and the G8. The latter argued that emission cutbacks could only be a global effort if it involved China and India as the two countries are in the midst of socioeconomic development and thus account for a large share of global emissions; see "G8, Emerging Nations Seek 'Deep Cuts' in CO_2 but Wrangle over Target—yodo", in: *BBC Monitoring Newsfile* (9 July 2008).

are operated globally. One of the most profound impacts of globalisation has been on the way that governments' roles are being perceived—the state in the era of globalisation is no longer prevalent in the provision of public goods. With the exception of such sacrosanct areas as national defence, for instance, most public services have been privatised based on the premise that competition in the free market would help provide public goods at a lower price. Energy supplies is one such area, where increased decentralisation of the energy market in countries worldwide aims to derive the best benefits for consumers through competition between energy suppliers.

However, the use of markets alone to provide energy security encompasses its own flaws, especially when the government is unwilling or unable to step into rectify negative externalities. There are examples where markets have failed to ensure energy security. For instance, the US$44 million Chalillo Dam project, which was initiated in the late 1990s in Belize, became a controversy over the lack of government oversight that resulted in environmental degradation and more costly instead of cheaper energy supplies for the country's citizens. In fact, based on an independent economic analysis, critics charged that the dam actually benefited Fortis Inc.—a Canadian company that owns 68 % of the stakes in the South American state-owned power supplier, Belize Electricity Limited—instead of the average Belizean.[6] Another notable example was that of Amnesty International accusing the American oil giant, ExxonMobil, in 2005 of prioritising profits over human rights in its involvement in the multi-billion Chad-Cameroon oil pipeline. The pipeline agreement was claimed to have been made without public purview, hence making oil companies de facto unaccountable for rights abuses in the pipeline zone.[7]

With the steady increase in energy prices since 2004, oil and gas are being viewed by the governments of some producer states as 'strategic commodities' that could be exploited for parochial purposes. Emergent 'resource nationalism' in some countries has even cast uncertainty over the supply of energy. For instance, since 2004, Hugo Chavez, the President of Venezuela, has renationalised the country's oil sector and forced foreign investors who refuse to relinquish the majority ownership stakes to accept higher tax payments or leave.[8] In a sign of deteriorating bilateral relations, President Chavez threatened to sever oil supplies to the US if the latter continued to harbour perceived aggressive intent towards his government.[9] More often than not, it is the average citizen—the ultimate end-user of energy supplies—

[6] Compared with the US$3.4 million in earnings from over 800 GW of energy sold in Newfoundland and Labrador, Canada, profits from the Chalillo Dam raked in US$5.6 million for Fortis Inc. even as Belizeans paid the company more than three times the average energy rates in Canada; see Loverock (2002, pp. 1–2).

[7] "Amnesty International Says US Consortium's African Oil Pipeline Threatens Human Rights", in: *Associated Press* (7 September 2005).

[8] "ANALYSIS—Oil Firms to Take Latest Chavez Bombshell in Stride", in: *Reuters* (10 January 2007).

[9] President Chavez announced that his government had uncovered a plot by active and retired Venezuelan military officers, with alleged US tacit approval, to stage a coup against his

who is the most affected in such situations. A case in point is that of Myanmar, where the ruling military junta has reportedly exploited the country's oil and gas revenues to fund grandiose projects, including substantial build-up of armaments for its armed forces, even as the population at large remains in abject poverty.[10]

The above examples underline the fact that reliance on markets or governance alone cannot ensure energy security and that market forces cannot be held to be fully responsible for the global energy system. There is a need for appropriate contingency measures in the event that markets fail to respond quickly to demand signals. Louder grassroots appeals calling for government interventions that can arrest prevalent energy problems and alleviate the plight of the man on the street point strongly to a need for stronger government roles in ensuring energy security. However, the geopolitics of energy, which involves both energy supply and access to energy resources, has traditionally been a driving force for global prosperity and security. Any disruption of energy supply could impinge on virtually every nation; energy crises could easily unravel democratic and authoritarian regimes in developed and developing countries alike. The political power of energy resources—with respect to political and industrial players as well as the common man—and the potential risk of its overuse or abuse cannot be underestimated, and that being so, the market domain will have to reconcile with legal and regulatory measures that make certain that governments continue to play a viable role in ensuring energy security at the national and societal levels. As such, a combined approach of markets and governance is required in the form of good market governance. The following section outlines the principles of good market governance prior to discussing its intricate links with energy security.

2.3.2 *Principles of Good Market Governance*

Good market governance is defined as "the best set of all laws, regulations, processes and practices that affect the functioning of a regulatory framework and the market" (Hancher et al. 2004). Market governance as a concept is not of an economic nature and, therefore, does not offer a framework for assessing the economic efficiency of a regulatory framework. It does, however, enable the legislature or the executive branch of a government as well as administrative authorities to formulate laws, policies, regulations and decisions of a high quality. No doubt, this applies greatly to the energy sector, which is sandwiched between the market domain and an acute need for a certain degree of government involvement. Good market governance is derived from documents reflecting the

(Footnote 9 continued)
government; see "Venezuela Expels US Ambassador, Threatens to Cut Oil", in: *Agence France Presse* (12 September 2008).

[10] "Activists Urge UN to Impose Energy Sanctions on Myanmar", in: *The Oil Daily* (21 November 2007).

best practices of market regulation in that the practices ensure that regulation promotes competition and welfare while not restricting economic growth (Hancher et al. 2004, p. 341). Good market governance, according to Hancher et al. (2004, pp. 342–352), should adhere to the following principles:

1. *Transparency*—This calls for the protection of all interested parties, including government, industrial and citizen players. It envisages all decisions and legislature to be made easily accessible, so that the state can act in an openly defensible manner while non-government parties are able to safeguard their rights. Legislature shall contain a clear formulation of the regulatory authority's powers and relation to purposes of law and also define clear areas of responsibilities that would be shared between authorities. At the implementation level, authorities would remain open to stakeholders about objectives, processes, records and decisions while being able to explain to the latter the rationale behind any decision made. In all, transparency would contribute to the legitimacy of the legal framework and actions of the agencies involved.
2. *Independent supervision of the market*—Stakeholders—political and industrial players, in particular—would not be able to unduly influence the outcome of regulatory procedures. However, independence does not equate to complete autonomy from government policies but rather means that the administrative authority is independent in implementing the regulations and policies free from government intrusion while, at the same time, heeding government policies as stipulated. With independent supervision of the market, issues with regard to regulations would be less likely to create controversies and hence less prone to allegations of partiality and arbitrariness.
3. *Clear legislative mandate*—This notion is strongly related to the principles of transparency; the independent authority would implement the will of the legislature as expressed in the legislative outcomes. The legislature would not only serve as a mandate for the authority of the administrative body charged with regulating the market but also provide clear guidance to the body on implementation issues.
4. *Flexible powers*—While the regulatory authority is supposed to follow the mandate of the legislature, it should not be placed in a strait-jacket, receiving well-delineated but limited powers to act. This is particularly so for such sectors as the energy and technology industries, where situations could evolve more quickly than the legislature can react. However, the risk of the agency being afforded too much leeway, allowing it to act beyond its established jurisdiction, also remain.
5. *Proportionality*—This principle means that regulatory action would only be taken when really necessary, that the measures chosen are appropriate to achieve their goals and that the effects of the measures chosen are proportionate to the objectives. As such, the principle of proportionality mainly protects regulated firms so that interference via regulation would be minimised, placing limits on the scope and substance of regulatory measures and hence forcing political and regulatory executives to follow transparent and efficient procedures, thereby also improving on accountability to the public and corporate sectors.

6. *Consistency*—Over time, consistency of treatment could serve as a means of improving confidence in the regulatory regime. This is linked to the provision of consistent and fair rules that do not adversely affect business performance of a specific participant in the regulated sector. Key mechanisms for achieving consistency include the formulation of clear and substantive rules and transparent procedures.
7. *Predictability*—By establishing decision-making criteria that are well defined, the provision of clear timetables for the review of standards and regulations as well as transparent decision-making procedures, the principle of predictability, if achieved, would allow regulated firms to plan with confidence for the future and be assured that their investments, many of which require long time horizons, will not be generally threatened by unexpected changes in the regulatory environment.
8. *Accountability*—Strongly associated with the principle of transparency, accountability would serve to protect the interests of all parties involved—regulated firms, citizens and the government itself—through a process of explanation, participation and control. Regulatory authorities would be held accountable to the political control instruments, to the judiciary through legal instruments, to citizens by explaining and publishing policies as well as to other interested parties, such as corporate firms, via public consultation procedures. However, for accountability to work, all non-governmental parties should be allowed to air their opinions on proposed measures taken by the regulatory authority through feedback. Judicial review of administrative actions taken by the regulatory authority would be an indispensable element to ensure accountability.
9. *Respect for general principles of competition policy*—The substance of regulatory measures should not conflict with the general competition law regime, as there could be risks of duplicity and rivalry if more than one agency is empowered to take action against certain behaviour of the regulated firm. Regulators should then be obliged to respect the general competition law and to consult competing authorities before taking action, thus preventing duplication of procedures and forum shopping.

2.4 Market Governance and Energy Security

2.4.1 Market Governance and Its Importance

It is necessary to note that the role of law in energy security is a double-edged sword. On the one hand, constitutional restrictions on private energy investments in Latin America, for instance, were able to stifle the energy market by precluding competition among industrial players. Conversely, legislation also helps ensure that citizens are rightfully entitled to adequate energy access and remain protected against errant profit-seeking energy firms. As Barton et al. (2004a, p. 463) highlight, countries vary considerably when it comes to the degree to which energy security is deemed as an explicit legal objective of decision-makers although, in the positive, general sense, the constitutional (and legislative) dimension helps protect energy infrastructure and supply against civil unrest while legally justifying government action to ensure energy security. Good market governance is thus necessary to formulate and implement effective energy policies that are underpinned by a combination of market forces and legal/regulatory framework.

The policies of consumers and producers, as a result of a dynamic energy market, may vary over time in response to each other or according to market developments. Elucidating this interaction, Birol points out that energy prices are likely to increase should producers' investments decrease (2007, pp. 14–15). In their defence, consumers could adopt policies that curb demand growth and import dependence by tempering the long-term effects of lower producer investments on prices and the amplification of depressive effects of increased prices on global demands. If successful, these policies would more likely lead to producers adopting measures to sustain production and global market share, thus reducing energy prices (Birol 2007). Since market forces cannot be held fully responsible for the global energy system, appropriate contingency measures would be required in the event of market inability to respond quickly enough to demand signals (Bochkarev and Austin 2007, pp. 6–8). Good market governance would thus be crucial in preventing producers and consumers from attempting to preserve their 'rights' to fluctuate demands arbitrarily outside legal frameworks, which could frustrate efforts to fund major energy investments and socioeconomic development projects.

Bochkarev and Austin (2007) suggest that establishment of the best national resilience and international contingency practices as well as international mechanisms that promote information sharing should be encouraged. At the national level, legal and regulatory measures emplaced by the government as an insurance against upheavals in the volatile and unpredictable global energy market would include, inter alia, price controls, restrictions on energy exports, subsidies as well as multinational frameworks to ensure the physical security of energy infrastructure, for instance, for combating unconventional maritime security threats, such as terrorism at sea. Indeed, fearing the politico-societal consequences of rising energy prices, some governments have resorted to subsidies although such moves could hasten demand growth by encouraging increased consumption and thus raise energy prices.[11] The long-term strategy would logically be to reduce energy consumption and promote the use of alternative energy sources.

---

[11] This is the case in many countries worldwide, where political security is deemed as top priority in the face of massive street protests over populist calls for the government to raise subsidies in order to alleviate increased energy prices. Such measures might invariably lead to economic slowdown; see "With Fuel Subsidies, Solutions Beget Problems: Unrest Quelled, but No Incentive to Save", in: *International Herald Tribune* (29 July 2008).

Good market governance is no less important from the international standpoint. For example, in July 2008, Russia sharply reduced oil supplies to the Czech Republic without any explicit reason, probably in response to the latter signing an agreement allowing the US to base a radar installation in the country as part of its National Missile Defence (NMD) programme, which the Russian government perceived as a threat to its national security.[12] From the consumer's point of view, the episode highlighted the importance of adequate regulatory mechanisms that could have provided for transparency on the security of supply, and thereby insulated the Russia-Czech oil supply from likely domestic political interference emanating from Moscow. The need for good market governance was also reflected in measures that were taken to bolster the European Community Commission's powers to act in a possible oil crisis in the wake of Iraqi aggression in 1990.[13] The moves were widely criticised for being proposed without consulting member countries, with fears that such moves could send panic signals to the energy market.[14] In this case, respect for accountability emerged as being necessary for long-term commitments towards investing in the capital-intensive fossil fuel industries (Hancher et al. 2004, pp. 343–345).

From the supply and environmental perspectives, the long-term solution to energy security would be to boost investments in both traditional and alternative energy sources, as satisfying the rising global demand for energy does call for corresponding investments into energy infrastructure. Pascual (2008) points out that, according to company plans, capital expenditure by leading energy companies increased sharply in nominal terms over the first half of the century's first decade and continued so until 2010. Even so, notwithstanding cumulative energy infrastructure investments that are projected to amount to about US$20 trillion (in 2005 dollars) over 2005–2030, there remain lingering uncertainties over the exact cost of discovering and exploiting energy sources over the coming decades.[15] What is more, the prospect of additional energy capacity generation from such increased investment spending is being blunted by rising costs. Not surprisingly then, investments in 2005 were reportedly lower than that in 2000 and capacity additions due to planned upstream investment up to 2010 were expected to boost global spare crude production only slightly (Pascual 2008, p. 6)—one of the

[12] Reports suggest that probably less than 300,000 metric tonnes of oil would be supplied instead of the 500,000 metric tonnes that was contractually agreed upon; see "Czechs Report 'Sharp' Cut in Russian Oil Supplies after Signing of US Radar Deal", in: *BBC Monitoring European* (14 July 2008).

[13] Member states would not easily agree to transfer control of their energy stocks to the Commission or empower it to dictate crisis measures for the entire regional body given the lack of an adequate framework that was based on general consensus; see "EC Energy Crisis Measures Ill-Timed, Diplomats Say", in: *Reuters* (25 October 1990).

[14] Ibid.

[15] Investments totalling over US$11 trillion are required in the power sector; capital expenditure is expected to be US$4.3 trillion and US$3.9 trillion in the oil and gas sectors, respectively. About half of all energy infrastructure investments will be in developing states, where demand and production are projected to increase the fastest. (Bochkarev and Austin 2007, p. 3).

reasons being regulatory delay—underscoring the fact that while securing reliable and affordable energy depends largely on adequate investments, this would only be possible in an environment of good market governance.

Problems remain with regard to the usage of alternative energy options. As the world population grows and demand for food increases, biofuel development using currently available technology would compete with food production for agricultural land and resources.[16] Where nuclear energy is concerned, nuclear power would become more viable only if governments are willing to facilitate investments and satisfactorily address concerns over plant safety, waste disposal and technology proliferation (Bochkarev and Austin 2007, pp. 10–11). Significant R&D is needed to mature such technologies. However, unlike established energy giants, small- and medium-sized enterprises (SMEs) that specialise in innovative energy solutions would likely be deterred by significant cost-benefit calculations. Indeed, the R&D of energy efficiency and alternative energy technologies involves long lead times and considerable costs, and such constraints call for prudence on the part of SMEs in particular. Here too, a business environment conducive for energy investment could be facilitated by the exercise of proper market governance.

Climate change adds another dimension to the dialogue on good market governance for energy security. In the face of threats to the ecosystem from climate change, environmentalists are prominently appealing for a reduction in GHG emissions. With the help of advanced media, their views could find resonance, especially in more educated and affluent societies, and potentially influence government policies. In 1991, for instance, environmentalists from Friends of the Earth challenged the Hong Kong government to develop an energy database and policy instead of establishing ad hoc measures only in times of panic, such as during the 1973 oil crisis.[17] To achieve climate change objectives while ensuring sustainable access to reliable, adequate and affordable energy sources, incentives must be created to drive changes in consumption patterns as well as R&D and dissemination of energy production technologies. This is a monumental task given that current policy and available technology will not be able to achieve the target of carbon dioxide emissions less than 450–550 parts per million by 2050 (Bochkarev and Austin 2007, pp. 9–12). An alternative here for the international community would be to price carbon in order to curb consumption, spur technological innovation, affect fuel choices and stimulate investments.

Where governments were previously primarily concerned with tackling the issues of energy supply and demand and the attendant effects on economic wellbeing on the whole, in the contemporary era, relying solely on either free market

[16] The Organisation for Economic Co-operation and Development (OECD) has refocused its efforts on reducing energy consumption in the transportation sector and to embark on second-generation biofuel research instead; see "Biofuels: OECD Report Blasts Biofuels as 'Costly and Ineffective'", in: *Europe Agriculture* (28 July 2008).

[17] "Energy Policy Branded as Pitiful by Friends of Earth", in: *South China Morning Post* (7 March 1991).

principles or government intervention could lead to energy insecurity. At present, energy (in)security is more than simply energy supply and demand, and policy-makers need to address a broader range of interrelated issues associated with energy use, including global warming. The interdependent nature of the global energy market, linking consumers and producers tightly together, necessitates a collective effort by all nations, especially where promoting energy efficiency and mitigation of climate change is concerned. Although progress in this respect has not been significant, examples can be found of the growing recognition that good market governance is key for ensuring energy security and of efforts towards finding a balance between market principles and regulation by the government. There is increasing worldwide interest in reinvigorating the governmental role, in conjunction with the market approach, for addressing issues relating to energy security. In February 2008, member countries of the European Commission, for instance, championed for the institution of an independent body, amalgamating the national energy regulators of member states, that would be equipped with flexible powers. The independent body would provide member representatives with powers to solve cross-border energy issues and allow them to take rapid and binding decisions when required in order to regulate the energy market.[18] Similarly, in 2006, calls for an independent power regulator were raised in Thailand prior to the privatisation of the Electrical Generating Authority of Thailand.[19] Likewise, in 2007, the United States Federal Energy Regulatory Commission (FERC) adopted new guidelines to enhance accountability among market operators and thus promote competition.[20] In short, good market governance, balancing the virtues of free market principles and government regulatory mechanisms, is increasingly being accepted as the way forward for mitigating externalities.

2.4.2 A Case Study: Japan

Respect for the principles of good market governance, according to Hancher et al. (2004, p. 341), would contribute to a good-functioning regulatory framework that will likely increase business confidence and produce stability required for stimulating long-term investment, so that the interests of consumers are better served with an appropriate supply and a dynamic market. To strengthen this argument, a case study of Japan is discussed here, which imparts valuable lessons on the importance of proper energy governance for energy security.

---

[18] "Legal Battle Brews over EC's Regulatory Agency Plan", in: *EU Energy*, 176 (8 February 2008).

[19] "Thailand: Democrats Criticize Government's Energy Policy", in: *Thai News Service* (11 August 2006).

[20] "Proposed Changes Would Raise Accountability of Market Operators", in: *Platts Commodity News* (21 June 2007).

2.4.2.1 Japan in the 1970s and Early 1990s

Japan witnessed a public hysteria to stockpile consumables during the first oil crisis in 1973 that served to further drive up energy prices. To an important extent, Japan's weakness in implementing an adjustment response at the time was attributed to factional politics and rigidly compartmentalised policymaking (Weatherford and Fukui 1989, pp. 605–608). Policy implementation was highly contingent and based on ad hoc measures, not to mention also the inconsistencies and incoherence in the policymaking process. Under demands for public–private cooperation and coordination in order to secure Japan's economic development and position, the Japanese government first established a Ministerial Council on General Energy Policy in 1975 to provide an energy security framework, which paved the way later in the year for the enactment of the Basic Direction on General Energy Policy. The latter could facilitate energy diversification and promote indigenous energy R&D and energy conservation. One of the most notable post-1973 initiatives, however, was a long-term policy, titled 'Energy Stabilization Policy for the Coming Decade: A Choice for State Supply', enacted in 1976 to provide energy projections through a diversity of openly accessible government and independent sources. This forecast system was especially crucial since it provided a price signalling system in a highly competitive energy market while serving as a platform for the Japanese government to draw attention to problem areas and opportunities, defend public resource allocation as well as trigger private sector initiatives (Nemetz et al. 1984–1985, p. 560).

As a result of these developments, a second oil crisis in 1979 saw a different outcome in Japan, where no public hysteria ensued. For instance, wholesale and retail price stability was restored by May 1980 while a 50 % increase in exports saw Japan's trade account surplus return by mid-1980 (Weatherford and Fukui 1989, p. 614). On analysing the effects that the oil crises of the 1970s had on advanced economies, Japan was found to have registered a higher economic growth in 1980 (at 5 % gross national product growth) than in 1974 when growth was negative (−0.5 %) (Weatherford and Fukui 1989, p. 615). In comparison with the US and UK, whose economic growths were −1.4 and −2.0 %, respectively, in 1974 and −0.7 and −2.3 %, respectively, in 1980, Japan's performance may have been due to its improved energy governance structures, put in place following the lessons learnt from the 1973 oil shock (Weatherford and Fukui 1989, p. 615).

The Japanese government played a less interventionist but more consistent approach during the second crisis—its key role was psychological, instilling preparedness and optimism among investors and consumers. Some of the principles of good market governance discussed earlier seem applicable in Japan's case. The diverse information provided by the forecast system, arguably one of the most important aspects of the post-1973 policies, may have helped in promoting predictability, transparency, independent supervision of the market and accountability. Consistency was ensured by the government's policy commitments, which were reinforced by the oil crisis of 1979. Proper market governance may have helped the Japanese economy absorb the second oil shock more effectively and rapidly than in 1973. Then, in 1990, during the US-led coalition's military build-up in response to Iraq's invasion of Kuwait, a potential energy crisis did loom large again for Japan. However, the Japanese society as a whole was better prepared, drawing on lessons learnt during the 1970s, and implemented contingency plans for energy savings.[21] On the supply side, Japanese energy industries remained unfazed with regard to energy infrastructure investments in the Middle East.[22]

2.4.2.2 The Fukushima Accident and Energy Governance in Japan: A Postscript

Proper energy governance in Japan has ensured energy security since the two oil shocks of the 1970s. However, the Fukushima accident in March 2011 could have exposed inadequacies in the pre-existing scope and extent of energy governance, which may have to be further ascertained in the ongoing government review of the latest disaster. While many reasons may have contributed to the accident, improper relationships between the regulators and those being regulated seem key among these, as they helped conceal the true technological and economic status of the nuclear power plants in question with regard to their operation and maintenance records—the nuclear power plants that were inundated by the tsunami and eventually malfunctioned would have been decommissioned by the time the tsunami struck if their technological status were reported and known to the regulator. Had such true records been revealed in the first place, the direct and indirect costs incurred following the nuclear accident could have been avoided.

The Fukushima accident has given Tokyo the opportunity, although rather forced, to review energy governance in Japan. This may propel the government to adopt market governance in the electricity industry, which might in turn lead to the demolition of the industry's existing region-based monopoly structure. The long-lasting effects of the nuclear accident may serve as a reminder for the government and the public alike of the real costs of twisted energy governance. Then again, it was functioning energy governance that eventually helped Japan overcome the apparent post-Fukushima energy crisis in the summer of 2011. Nonetheless, some loopholes might have existed and Tokyo is in the process of reviewing what went

---

[21] "Gov't Calls for Energy Savings, Cites Gulf Crisis", in: *Associated Press* (13 August 1990). In the US, meanwhile, the gulf crisis created public panic over the possibility of an energy crisis, with some blaming the surge in gasoline prices on energy firms' greed; see "Half in US Fear Gas Crisis as a Result of Middle East Conflict", in: *Associated Press* (13 August 1990).

[22] The export of petrochemical plant technology to the Middle East registered a significant surge (48.9 %) between April–September 1990, notwithstanding the Iraqi invasion and subsequent allied military build-up, compared to the same period in 1989. Even so, industrial officials admitted that it would have been almost impossible to resume plant construction in the region, if the war expanded, due to physical security concerns; see "Japanese Industry Shocked, but Copes Coolly with War (Iraq-Kuwait Crisis, 1990)" in: *Japanese Economic Newswire* (17 January 1991).

right and wrong in Japan's energy governance and will undoubtedly in the future produce a better variant of such structures to cope with future challenges. The Fukushima accident, in sum, highlighted the importance of adequate energy governance—addressing the adequacy of supply as well as associated environmental and human costs—for ensuring energy security when using nuclear power.

2.5 Conclusion

This chapter explores what proper market governance is and how it can be combined with government regulatory measures and free market principles to ensure energy security that is defined as 'an adequate and reliable supply of energy resources at reasonable prices'. Energy security is a perennial issue not only for energy-consuming countries but also for self-sufficient energy-producing ones. The projected increase in demand for energy, especially in rapidly growing economies such as China and India, as well as the continued reliance on fossil fuels in the foreseeable future will place strains not just on the future of supplies but also the environment. As such, energy security is no longer merely a concern over the security of supply and demand; it is also related to environmental impacts caused by climate change from prevalent energy usage patterns. To meet the projected increase in energy demand, investments in energy infrastructure, including an overhaul of the production facilities for fossil fuels and R&D into cleaner alternative sources of energy, are crucial. However, the modernisation of energy infrastructure will take time and colossal amounts of funding, the latter calling for substantial investments from the private sector. The business climate for the energy market thus needs to remain investor-friendly in order to encourage and sustain financial commitment from energy investors. As discussions have shown, instances of alleged profiteering by energy companies, such as Fortis Inc., and the overuse of governance, as in the case of Venezuela, indicate that using either the market or governance approach alone may not foster such investments. In order to strike a balance between the market and regulatory domain, good market governance is essential.

The importance of exercising proper and adequate market governance is highlighted by Japan's responses to the oil crises of 1973 and 1979. In contrast to the shocks experienced by the Japanese public during the first oil crisis due to its heavy reliance on the free market and inadequate government intervention, Japan could weather the second crisis better as a result of the formulation and implementation of policies that were able to instil confidence in investors and consumers alike. Similarly, the Fukushima accident also underscores the importance of properly functioning governance, balancing regulatory measures and market mechanisms, for ensuring energy security.

The eventual goal of government intervention would be the formulation of a standardised set of principles for good market governance that draws from the best national or regional practices. Until such a time, every player in the global energy market would have to remain on guard against future energy crises, which are by no means remote possibilities. In the short run, proper market governance could probably facilitate investments needed for modernising and expanding production capabilities of traditional energy sources and help to meet the rising global energy demand. However, market governance would be even more crucial in the long term, facilitating investments in energy-efficient and alternative (cleaner) energy sources that will not only stabilise this demand but also mitigate global warming, which involves dire consequences for the environment and ecosystem.

The time is right for a reassessment of the scope and roles of market and governance for ensuring energy security. The following policy recommendations, if further developed, point to areas for potential research or policy action: (1) sociopolitical stability that provides an environment conducive for foreign investments; (2) clearly defined national energy policies and regulations instead of ad hoc reactive measures in times of market contingencies; and (3) promoting dialogue between governments, industrial players and civil societies in order to raise awareness on the importance of good market governance. Energy security through the exercise of good market governance would instil producer–consumer confidence and facilitate energy-related investments in the dynamic and uncertain global energy market, today and tomorrow alike.

References

Barton B, Redgwell C, Ronne A, Zillman DN (2004a) Energy security in the twenty-first century. In: Barton B, Redgwell C, Ronne A, Zillman DN (eds) Energy security: managing risk in a dynamic legal and regulatory environment. Oxford University Press, New York

Barton B, Redgwell C, Ronne A, Zillman DN (2004b) Introduction. In: Barton B, Redgwell C, Ronne A, Zillman DN (eds) Energy security: managing risk in a dynamic legal and regulatory environment. Oxford University Press, New York

Birol F (2007) Energy security: investment or insecurity. Coping with crisis, CWC Working paper series, International Peace Academy

Bochkarev D, Austin G (2007) Energy sovereignty and security: restoring confidence in a cooperative international system, Policy Pap, 1:1–33 (East-West Institute)

Hancher L, Larouche P, Lavrijssen S (2004) Principles of good market governance. Tijdschrift voor Economie en Management 49(2):339–374

IEA (International Energy Agency) (2008a) Oil market report: a monthly oil market and stocks assessment, July 10. IEA, Paris

IEA (International Energy Agency) (2008b) Worldwide trends in energy use and efficiency: key insights from IEA indicator analysis. IEA, Paris

Loverock K (2002) Chalillo chill: critics slam the dam that a Canadian company wants to build in Belize's valley of the Scarlet Macaw (sustaining livelihoods). Altern J 28(2):1–2 (22 March)

Nemetz PN, Vertinsky I, Vertinsky P (1984–1985) Japan's energy strategy at the crossroads. Pacific Aff 57(4):553–576

Noe SY, Pring G (2004) 'The 'fear factor': why we should not allow energy rhetoric to trump sustainable development. In: Barton B, Redgwell C, Ronne A, Zillman DN (eds) Energy security: managing risk in a dynamic legal and regulatory environment. Oxford University Press, New York

Pascual C (2008) The geopolitics of energy: from security to survival. Brookings Institution, pp 1–16
Verrastro F, Ladislaw S (2007) Providing energy security in an interdependent world. Wash Q 30(4):95–104
Weatherford MS, Fukui H (1989) Domestic adjustment to international shocks in Japan and the United States. Int Organ 43(4):585–623

Abbreviations

FERC	United States Federal Energy Regulatory Commission
GHG	Greenhouse gas
GW	Gigawatt
HSS	School of Humanities and Social Sciences
IEA	International Energy Agency
IPCC	Intergovernmental Panel on Climate Change
mmbd	Million barrels per day
NEA	National Environment Agency, Singapore
NMD	National Missile Defence
NTS	Non-traditional security
NTU	Nanyang Technological University
OECD	Organisation for Economic Co-operation and Development
R&D	Research and development
RSIS	S. Rajaratnam School of International Studies
SLOC	Sea lines of communication
SME	Small- and medium-sized enterprise
UK	United Kingdom
US	United States

Author Biographies

Youngho Chang (Singapore): Assistant professor of Economics at the Division of Economics, Nanyang Technological University (NTU), Singapore. Since 1999, he has taught resource and energy economics, environmental economics and macroeconomics at the National University of Singapore and NTU. He has also contributed to various academic journals, including *Econometric Theory*, *Energy Policy*, *Economics Letters*, *International Journal of Global Energy Issues* and *International Journal of Electronic Business Management*. He received a BASc in landscape architecture from Seoul National University, South Korea; an MA in economics from Yonsei University, South Korea; and a PhD, also in economics, from the University of Hawaii at Manoa, US.

Swee Lean Collin Koh (Singapore): Associate research fellow at the S. Rajaratnam School of International Studies (RSIS). He was a research analyst on energy security at the RSIS Centre for Non-Traditional Security (NTS) Studies from 2008 to 2010, during which time he contributed to the centre's in-house publications, primarily focusing on energy security. He is also co-editor of a book (with Basrur, R.): *Nuclear Power and Energy Security in Asia* (Routledge, forthcoming in May 2012).

Chapter 3
Energy Security and Climate Change in ASEAN: Implications and Policies

Youngho Chang and Lixia Yao

Abstract This chapter presents a quantifiable definition of energy security and the various indicators generally used to measure it. These indicators are then applied to Association of Southeast Asian Nations (ASEAN) countries in order to evaluate their past energy security policies and activities, which include cross-border infrastructure building and hydropower exploitation projects in the region. ASEAN's efforts at enhancing energy security are analysed relative to its efforts for climate change mitigation, as energy security and climate change have profound impacts on each other. The chapter also suggests alternative approaches for addressing issues related to energy security and climate change in the region.

Keywords ARF debates · ASEAN · Climate change · Energy diversity · Energy security · Hydropower exploitation · Infrastructure building

---

Y. Chang (✉)
HSS-04-65, Division of Economics, School of Humanities and Social Sciences (HSS), Nanyang Technological University (NTU), 14 Nanyang Drive, Singapore 637332, Singapore
e-mail: isyhchang@ntu.edu.sg
http://econ.hss.ntu.edu.sg/Pages/Home.aspx

L. Yao
S. Rajaratnam School of International Studies (RSIS), Nanyang Technological University (NTU), Block S4, Level B4, Nanyang Avenue, Singapore 639798, Singapore
e-mail: yaol0007@e.ntu.edu.sg
www.rsis.edu.sg/nts

M. Caballero-Anthony et al. (eds.), *Energy and Non-Traditional Security (NTS) in Asia*, SpringerBriefs in Environment Security, Development and Peace 1,
DOI: 10.1007/978-3-642-29706-9_3,

3.1 Introduction

Energy security and relevant policies within the Association of Southeast Asian Nations (ASEAN) region are discussed at length in this chapter. To explain the notion of energy security in ASEAN, the definition of energy diversity and its measurements were adopted and modified, following which its alternative definitions and measurements are described. Energy security policies in the region are presented, with a systematic overview of past ASEAN energy security activities and projects as well as present policy debates, as witnessed during the recent ASEAN Regional Forum (ARF) meetings on climate change and security in Cambodia and Brussels. Section 3.2 of the chapter introduces the theoretical frameworks and indicators of energy security. Section 3.3 discusses ASEAN's policies and initiatives with regard to climate change and energy security. Finally, Sect. 3.4 provides a brief summary of key findings.

3.2 Energy Security: Concept, Measurements and Indicators

3.2.1 Energy Security: Concepts and Measurements

Energy security is a concept of multiple dimensions that assumes different specificities depending on the country, energy source and timeframe (Chester 2008). There is no universally accepted or unified definition for energy security although several studies have tried to provide a systematic and comprehensive definition for it and a variety of definitions with differing components have been proposed by many such efforts. Generally speaking, energy security deals with security of all kinds of energy resources and covers several elements and factors. Energy security is often described as the availability of stable and reliable energy supplies at all times in various forms in sufficient quantities and at reasonable and/or affordable prices (Yergin 1988; CIEP 2004; IEA 2007b; Chang and Lee 2008). However, no energy system, irrespective of whether for the short or long term, can be entirely secure. Reflecting this, energy security is even perceived as "a problem of risk management" that reduces "to an acceptable level the risks and consequences of disruptions and adverse long-term market trends" (IEA 2007b, p. 161).

Some definitions stress on the sustainability factor when interpreting energy security, defining it in terms of a country's ability to guarantee the availability of energy resource supply in a sustainable manner, while also emphasising that energy price shall be reasonable (APERC 2007). In keeping with this line of thought, energy security is attributed with three key elements—adequacy, reliability and a reasonable price—and defined as having adequate resources and reliable supply of energy at a reasonable price. Here, a reasonable price refers to one that is devoid of the exercise of excessive market power and does not adversely affect the economic performance of a country (APERC 2007; Chang and

Lee 2008). A country's efforts to guarantee energy security, according to this approach, are normally focused on ensuring three elements—adequate resources, reliable supply and reasonable price.

Several alternative concepts and measurements of energy security also exist. An increasingly popular view in the literature is that energy security is an issue so complex that a holistic approach is needed to address it and "capture the complexity of the concept" (Sovacool and Mukherjee 2011, p. 5346). Several 'holistic notions' have been proposed to this end. One—the 'five Ss'—describes energy security as a situation where five characteristics dominate the whole energy system: surety, survivability, supply, sufficiency and sustainability (Kleber 2009).[1] Alternative notions, such as the multiple 'As'—energy resource availability, accessibility barriers, environmental acceptability and investment cost affordability (APERC 2007); or, availability, adequacy of capacity, affordability and sustainability (Chester 2008)—have also been put forth. Meanwhile, Sovacool and Brown argue that "energy security should be based on the interconnected factors of availability, affordability, efficiency, and environmental stewardship" (2010, p. 81). In a later work, additional dimensions of 'regulation and governance' were also included, so that energy security involved availability, affordability, technology and development efficiency, environmental and social sustainability, and regulation and governance (Sovacool and Mukherjee 2011). Sovacool et al. subsequently allocated a total of 20 indicators for measuring national energy security performance using this five-dimensional concept of energy security (Sovacool et al. 2011). The authors used the new definition to assess the energy security status of various countries and found that the energy security performance of the ten ASEAN countries had decidedly worsened during the 1990–2010 period. Brunei was the best performer among ASEAN countries, while Vietnam and Myanmar were the worst.[2]

3.2.1.1 Energy Diversity and the Status of ASEAN Countries

Energy diversity, a measurement indicator that reflects one aspect of reliable supply, is defined as "an evenly balanced reliance on a variety of mutually disparate options" (Stirling 2010, p. 1625). Energy diversity—similar to the concept of diversity employed in other disciplines, such as ecology, economics, taxonomy

---

[1] By this description, surety is "[a] condition which provides access to energy and fuel sources"; survivability means "energy and fuel sources are resilient and durable in the face of potential damage"; supply suggests "[an] identified and available source of energy, whether it is traditional fossil fuels, alternative energy or renewable energy"; sufficiency means "an adequate quantity of power and fuel from a variety of sources"; and sustainability requires operating practices to "be perpetuated by limiting demand, reducing waste and effectively utilizing alternative energy and renewable resources to the maximum extent possible" (Kleber 2009).

[2] Conversely, data from Table 3.1 suggests that Brunei's energy security status is relatively low mainly due to its high dependence on fossil fuels.

and so on—displays three basic properties: variety, balance and disparity (Stirling 1994; Grubb et al. 2006). A country may have as many energy resource options as it has indigenous energy supply sources, be it in the form of coal, oil, natural gas, renewables or the like. That is, variety is the answer to the question on how many options a country has with regard to energy resources. By the same yardstick, balance answers the question on how much a country relies on each energy source. Balance may be represented by a set of positive fractions for each energy source to the total energy supply, provided that all fractions add up to one in total (Stirling 2010, p. 1625). Disparity is defined as the manner and degree to which energy sources are distinguishable (Runnegar 1987). It is the answer to the question on how different a country's energy sources are from each other (Stirling 2010, p. 1626). By this definition, greater the variety of the types of a country's energy sources—together with a more even balance across energy sources and more disparate energy options—greater is the diversity of the country's energy system, and consequently more energy secure is the country in terms of its diversified energy supply.

Simple and straightforward measures therefore that indicate a country's energy security—such as the inverse of the number of energy resources used, the share of the most utilised resource, the share of fossil fuels used and the share of top five most utilised resources—could be established to examine its energy security status. All together, such measures help to examine the diversity of a country's energy resources and the degree of its dependence on fossil fuels.

For instance, the inverse of the number of energy resources, (1/n), reflects the variety aspect of energy diversity and is an overall indicator that is simple and straightforward. Here 'n' is the number of resources that a country uses—the bigger the number, the more the kinds of resources that are used in the country. A lower value for the indicator signifies a relatively low dependence on one fuel for a country, which also means that supply disruption of a fuel will likely cause lesser damage than otherwise and that the country is more energy secure. That is, a low '1/n' represents a relatively better energy security status for the country. Similarly, the share of the most utilised resource, (%), shows the degree of dominance or concentration of one kind of energy resource in a country. Thus, for a country, a lower number signifies lesser dependence of one energy resource and better status of energy security. This indicator reflects the balance aspect of energy diversity. The share of fossil fuels used, (%), shows a country's dependence on fossil fuels. Here too, a lower number means less dependence on fossil fuels and more usage of non-fossil fuels, and that the country is more disparate in terms of energy resources and, therefore, has a better status of energy security. The share of top five most utilised resources, (%), is an equally useful indicator since the top five resources must contain non-fossil fuels in the country's energy or fuel mix and, therefore, reflects its true and relative energy diversity. As in the others, a lower number indicates a higher level of diversity into non-fossil energy resources and thus that the country has a higher status of energy security (Chang and Yao 2011).

Table 3.1 Energy diversity of ASEAN countries, 2009

Country	Number of resources (inverse of the number of energy resources [1/n])	Most utilised resource (share of the most utilised resource [%])	Share of fossil fuels used (%)	Share of top five most utilised resources (%)
Brunei	0.5 (2)	Natural gas (73)	100.0	100.0
Cambodia	1.0 (1)	Oil (100)	100.0	100.0
Indonesia	0.11 (9)	Oil (47)	97.7	99.6
Laos	0.33 (3)	Oil (100)	100.0	100.0
Malaysia	0.13 (8)	Natural gas (51)	97.5	100.0
Myanmar	0.5 (2)	Natural gas (56)	100.0	100.0
Philippines	0.13 (8)	Oil (58)	92.5	100.0
Singapore	0.33 (3)	Oil (88)	100.0	100.0
Thailand	0.13 (8)	Oil (52)	96.6	98.1
Vietnam	0.17 (6)	Oil (37)	100.0	100.0

Abbreviations ASEAN = Association of Southeast Asian Nations; *n* = number of resources
Changes of values after 2009 are very small or nil
Source Chang 2009, p. 105. Copyright: World Scientific

The energy diversity of ASEAN countries was calculated and their energy security status evaluated using these four indicators. Table 3.1, which shows the energy security indicators and related energy security status of these nations in 2009, reveals that most ASEAN countries dominantly use one fuel—either oil or natural gas—and that, given such high dependence on fossil fuels, the energy security status of ASEAN is not good. However, ASEAN has been trying to improve its energy security status and has huge potential where renewable energy resources are concerned. Therefore, it could be expected that energy diversity and the energy security status of ASEAN will largely improve within the next decade.

Energy diversity is not only about the variety of energy resources though—a variety of energy infrastructure can also serve to diversify the sources and routes of energy supply. For instance, the new liquefied natural gas (LNG) terminal in Singapore will diversify its energy sources and energy supply routes. In Singapore, about 80% of electricity is generated from natural gas that is imported from Malaysia and Indonesia via pipelines. It is estimated that Singapore's demand for gas will exceed supply in the future. To increase and diversify the sources and routes of energy supply, in January 2005, the Singapore Energy Market Authority (EMA) undertook a feasibility study in relation to LNG import into Singapore. In 2006, the Singapore government announced its plan to build the country's first LNG receiving terminal to diversify its sources of natural gas and to meet the country's rising future demand for energy. The LNG terminal is targeted to be ready for start-up by 2013.[3] Figure 3.1 depicts the fuel supply options for Singapore and the countries from which it could source gas when the LNG terminal is complete.

[3] EMA, "Liquefied Natural Gas", at: http://www.ema.gov.sg/LNG/ (20 January 2012).

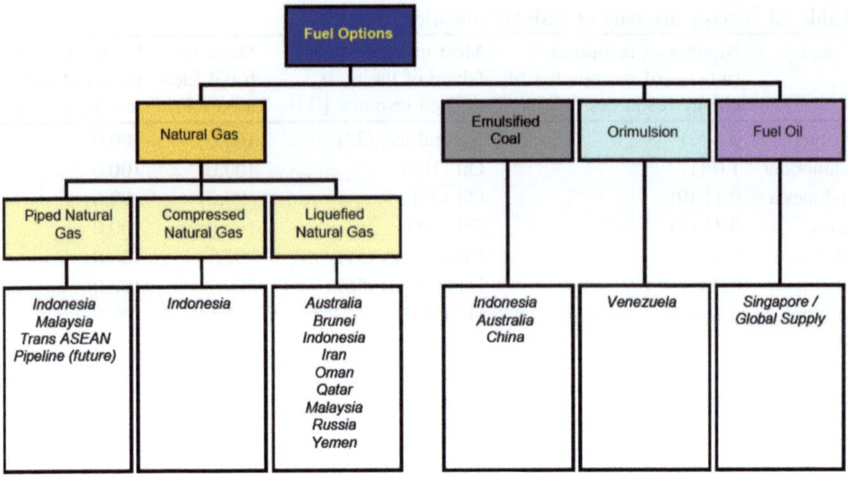

Fig. 3.1 Fuel supply options for Singapore. *Source* EMA 2005, p. 9

3.3 Challenges Facing ASEAN Countries: Climate Change and Energy Security

3.3.1 Climate Change Versus Energy Security

Climate change has generally caused an increase in global sea levels by 1.7–1.8 mm annually over the last century. The rate of rise in sea levels, however, has accelerated during the last decade to about 3 mm a year (ADB 2008b, p. 2). While most countries worldwide are bearing the brunt of climate change, those in Southeast Asia are likely to be particularly challenged by its effects. The Greater Mekong Subregion (GMS) is especially at risk due to its extensive coastlines and major deltas that are barely above the mean sea level (WWF 2009, p. 8). Both the dry and wet seasons are being exacerbated by climate change—increased glacial melt from the Himalayas will amplify the risk of flooding even as drought will be made worse by higher evaporation due to regional warming. The Mekong water flows will become less predictable due to climate variability, particularly impacting downstream countries, such as Cambodia, Laos and Vietnam. For instance, as a major food producer in the region, Vietnam is projected to be severely affected by a deteriorating ecosystem due to climate change since half of Vietnam's rice production is concentrated in the Mekong delta (WWF 2009, p. 8).

Climate change deliberations often arouse concerns over the future supplies of energy, as much of the increase in greenhouse gas (GHG) emissions is attributed to fossil fuel use. It is even possible that recommendations for climate change mitigation may compete with proposals for ensuring energy security and that policies that address energy security concerns may have significant implications for climate change mitigation efforts, just as both may directly affect changes in a country's fuel mix. For instance, climate change mitigation calls for GHG emissions to be reduced to a fraction of current levels. However, such reduction in emissions is likely to require countries to switch from their predominantly fossil fuel-based energy supplies to carbon-free energy sources, such as nuclear energy and renewable energy, which countries may find hard to reconcile with their pressing energy security concerns. Also, renewable energy and/or less carbon-intensive energy sources, such as natural gas, are likely to be a limited and costly option for most countries.

The climate change mitigation concerns of Southeast Asia too are in many ways contrary to its aspirations of regional energy security. For instance, the construction of dams on rivers will negatively affect the ecosystems alongside the river. Even so, not building dams for hydropower plants that could help to address the region's energy insufficiency may exacerbate its energy shortages. Indeed, the interconnectedness of the region's ecological, food and energy crises might stall and even reverse development in Southeast Asia. Not surprisingly, then, the interactions between climate change mitigation and energy security policies are attracting growing attention worldwide. Reflecting this trend, the national energy policies of countries, such as Australia, the UK and France, have put much emphasis on the linkages between energy security and climate change (IEA 2007a, p. 24). ASEAN too requires coordinated regional coping strategies and not individual or independent action from member countries, whether for adapting to climate change or for ensuring food or energy security, as the latter will likely have a negative impact on its neighbouring countries.

Most discussions on energy security suggest that a diversification of energy sources would contribute to a country's energy security. Much the same, the key recommendation, where climate change mitigation is also concerned, is for countries to diversify into energy sources that are less polluting in terms of emissions and not fossil fuel-based. Certainly, the way forward for negotiations on climate change mitigation and energy security concerns is likely to lie in efforts that aim to reduce the use of fossil fuels and diversify energy sources simultaneously (Gradziuk 2009, p. 28).

3.3.2 *ASEAN Initiatives vis-à-vis Climate Change*

ASEAN countries have initiated several measures to deal with the issue of climate change, which may be impacting energy security in the region in positive and negative ways. This section outlines the policies and activities that ASEAN has adopted in response to the challenges facing the region and discusses the ARF debates, which have been instrumental in recognising that although climate change presents countries with non-traditional security (NTS) threats, it is also a catalyst for international cooperation. Two recent ARF debates on the issue of climate change and its impacts on national security, particularly energy security, are highlighted.

3.3.2.1 Policies and Initiatives

In ASEAN countries, the combustion of fossil fuels and biomass in the industry, transport and agriculture sectors as well as in households releases large amounts of environmental pollutants. To alleviate this pollution, all ASEAN countries, except for Brunei, have ratified and acceded to the Kyoto Protocol, which aims to reduce GHG emissions and address climate change. ASEAN countries actively participate in the Clean Development Mechanism under the Kyoto Protocol, according to which developed countries invest in clean energy projects in developing countries to reduce GHG emissions. More importantly, during the ASEAN Summit held in Singapore in November 2007, two forward-looking documents[4] were signed to reaffirm ASEAN's commitment to mitigate the negative impacts of climate change. The first document enlists ASEAN initiatives to tackle regional environmental challenges and climate change. The second document pledges ASEAN's support for the United Nations Framework Convention on Climate Change (UNFCCC) and the Kyoto Protocol (Glover and Lee 2008, p. 3). In addition to intergovernmental declarations, specific air quality standards and monitoring are in place in most ASEAN countries; these countries have also taken steps to ensure that fuel combustion in the industrial and transport sectors does not release excessive amounts of environmental pollutants into the air (Glover and Lee 2008, p. 4).

3.3.2.2 The ARF Debates

To deal with climate change issues, two relevant debates were held at ARF seminars recently. In March 2009, the seminar on *International Security Implications of Climate-Related Events and Trends*, which was co-chaired by the EU and Cambodia, was held in Phnom Penh. It was agreed in the seminar that climate change presents non-traditional threats to national and regional security and that existing security instruments were not sound enough or sufficient to deal with these threats. However, it was also agreed that despite climate change being a 'threat multiplier', it could also be a 'peace multiplier' by way of catalysing regional and international cooperation. A need to develop regional scenarios and studies that could be an input for further deliberations on how climate change could become a

[4] ASEAN, "ASEAN Declaration on Environmental Sustainability", at: http://www.aseansec.org/21060.htm (27 January 2012); ASEAN, "ASEAN Declaration on the 13th Session of the Conference of Parties (COP) to the UN Framework Convention on Climate Change (UNFCCC) and the 3rd Session of the Conference of the Parties Serving as the Meeting of the Parties (CMP) to the Kyoto Protocol", at: http://www.asean.org/21078.htm (27 January 2012).

'peace multiplier' was highlighted. The main emphasis of the studies would be to develop suggestions and recommendations for further discussions.[5]

Another ARF seminar that witnessed a climate change debate was that held in Brussels in November 2010. Titled *International Security Implications of Climate Change*, the seminar was a follow-up to the one in Cambodia. It discussed the impact of climate change on various aspects of national security, including food security, energy security, etc., and reaffirmed that climate change was a 'threat multiplier' that could create or exacerbate insecurities and tensions both within and outside a country.[6] Mitigation was noted as the most important strategy for addressing climate change, meaning that countries should incorporate mitigation strategies into national development planning. It was emphasised that a regional approach was needed to further explore responses to the security risks associated with climate change.

The Brussels seminar concluded and recommended approaches and actions for addressing climate change. First, it pointed out that, in addition to efforts at the national level, regional political dialogue to promote understanding of the complex interlinkages between climate change and security implications needed to continue. Second, due to resource pressures and the complex and transboundary impacts of climate change on agricultural productivity, regional countries needed to pay attention especially to strengthening national and regional capacities on disaster risk reduction and management via development assistance and technology transfer, and so on. It was also highlighted that since there was no 'one size fits all' solution to risk reduction, the specificities of regions needed to be identified and taken into account. Third, effective collaborative platforms where organisations and authorities at all levels are engaged in promoting an understanding of the linkages between climate change and security needed to be set up. Fourth, scientific research was stressed. The setting up of a collaborative research project involving universities and research centres from the ASEAN and the EU was suggested to enhance the knowledge base of the impacts of climate change on security. Last but not the least, it was urged that regional initiatives, such as the Mekong river development projects, be utilised as possible approaches to address the issue of climate change and its impacts on security.[7]

---

[5] "Co-Chairs' Summary Report of the Seminar on "International Security Implications of Climate-Related Events and Trends"", Phnom Penh, Cambodia, 19 March 2009, at: http://aseanregionalforum.asean.org/files/library/ARF%20Chairman's%20Statements%20and%20Reports/The%20Sixteenth%20ASEAN%20Regional%20Forum,%202008-2009/Co-Chairs%20Summary%20of%20Seminar%20SI&CC.pdf (20 January 2012): 1–5.

[6] "Report on the Implementation of the European Security Strategy: Providing Security in a Changing World", Brussels, 11 December 2008, at: http://www.consilium.europa.eu/ueDocs/cms_Data/docs/pressdata/EN/reports/104630.pdf (20 January 2012): 5–6.

[7] "Co-Chairs' Summary Report of the Seminar on "International Security Implications of Climate Change"", Brussels, 18–19 November 2010, at: http://aseanregionalforum.asean.org/files/library/ARF%20Chairman's%20Statements%20and%20Reports/The%20Eighteenth%20ASEAN%20Regional%20Forum,%202010-2011/17%20-%20Co-Chairs%20Summary%20Report%20of%20ARF%20Seminar%20on%20Climate%20Change%20as%20of%20101129-1.pdf (20 January 2012): 1–5.

3.3.3 ASEAN Initiatives vis-à-vis Energy Security

ASEAN is not only highly dependent on fossil fuels but also has to deal with an indigenous fossil fuel supply that does not sufficiently meet its demands. In the oil sector, for instance, ASEAN's oil consumption, which was 179 million tonnes of oil equivalent (mtoe) in 2007, was projected to reach 267 mtoe by 2030 according to the International Energy Agency (IEA) (IEA 2009, p. 544). In contrast, the region's oil production is projected to decline steadily in the coming years, with estimates suggesting production of 2.4 million barrels per day (mmbd) in 2015 and 1.4 mmbd in 2030. Similarly, in the natural gas sector too, exports are expected to decline unless new gas fields are discovered. Some nations are also expected to increasingly import LNG (IEA 2009, p. 548). As a hedge against such risks and to improve its energy security, ASEAN has launched a series of initiatives that endeavour to make better use of its indigenous energy resources, including fossil fuels and renewable energy sources. These include, but are not limited to, the ASEAN Power Grid (APG), Trans ASEAN Gas Pipeline (TAGP) and energy development and cooperation in the GMS.

As ASEAN is rich in energy resources, it is feasible for its member countries to coordinate and establish a pan-ASEAN energy market that is economically sound and technologically viable. This is proved by ASEAN's plans for the establishment of the APG[8] for which the power authorities of ASEAN countries have agreed to interconnect electricity networks. In July 2003, the ASEAN Interconnection Master Plan Study (AIMS), which was a study on the APG, was approved. The plan selected 11 power grid projects for implementation. A subsequent memorandum of understanding (MOU) affirmed the benefits of the APG.[9] The APG interconnections will greatly add to the capacity of the power lines throughout the ASEAN region as they cover all ASEAN countries, from Yangon and Hanoi in the north, through eastern Malaysia and the middle part of the Philippines, to islands in southern Indonesia. The interconnections are, however, most concentrated between Thailand and Laos, which have active power trading agreements with each other.

Cooperation among ASEAN countries is not limited to the power sector. Blessed with rich natural gas resources, ASEAN member countries have also formulated a master plan that addresses issues relating to the commercial, legal and institutional aspects of the construction of the TAGP[10] project. ASEAN

---

[8] For the grid network, see "ASEAN Power Grid", at: http://www.siew.sg/sites/default/files/apg_only.jpg (8 February 2012); for the current status of APG, see Hermawanto, Bambang, "Report of the 8th Meeting of APGCC", Presentation at the 27th Meeting of HAPUA Council, Danang, Vietnam, 23 June 2011, at: http://www.hapuasecretariat.org/doc2011/Report%208_APGCC.pdf (8 February 2012): 3.

[9] ASEAN, "Memorandum of Understanding on the ASEAN Power Grid", at: http://www.aseansec.org/20918.htm (8 February 2012).

[10] ASCOPE, "Trans ASEAN Gas Pipeline Project (TAGP)", at: http://ascope.org/component/content/article/6-projects/28-tagp.html (8 February 2012).

countries signed the ASEAN MOU on TAGP[11] in July 2002 to facilitate its construction and assure long-term reliable gas supply. The routes for the TAGP will be similar to those of the APG, with a key difference being that the gas pipeline grid will not cover most parts of the Philippines although natural gas is increasingly an important energy source for the country. The TAGP, when completed, will dramatically reduce the region's dependence on imported energy, thus strengthening the region's energy security, and also helping to settle to some extent its emissions problems. An integrated regional energy market through APG and TAGP will not only invigorate the region by making energy resources in the region freely flowing but also strengthen mutual trust and confidence between members.

Apart from infrastructure building to further exploit the region's fossil fuel potential, ASEAN is also committed to developing its huge renewable energy potential, as seen from its plans for hydropower exploration. Although ASEAN has large potential in hydropower, its utilisation is very low. To further explore this potential, a technical assistance programme was initiated in 1992 by the Asian Development Bank (ADB) to promote subregional cooperation in hydropower development in the GMS region, especially among Cambodia, China, Laos, Myanmar, Thailand and Vietnam (Chander 2000, p. 3). The initiative also opens up possibilities for increasingly enhanced cooperation and integration in the GMS.

Cooperation within the GMS can be traced back to 1957 when the Mekong Committee was established. Its goals were largely to address the pressing issues of political instability along the river basin and to promote peace and prosperity through the joint use of the resources in the region (Hiroshi 2000). On the energy side, although the GMS was abundant in energy resources, large-scale energy development and energy cooperation did not begin until the 1990s, when the ADB project was launched. Three years after the ADB programme, an official Electric Power Forum was established as an advisory body to a newly inaugurated annual GMS Ministerial Conference in Yangoon. Cooperation among members has steadily increased since 1995 and several subregional hydropower projects have been developed, including the Lao PDR-based Theun Hinboun, Houay Ho and Nam Leuk hydropower projects that currently export electricity to Thailand. Other major existing and forthcoming hydropower export projects in the GMS include the Nam Ngum 2, Nam Theun 2 (NT2), Xe Pian-Xe Namnoy and Xe Khaman 1 in Laos; the Tasang in Myanmar; the Jinghong and Nuozhadu in China; and the Sambor CPEC (Cambodia Petroleum Exploration Co.) in Cambodia (APERC 2004, p. 13).

Existing GMS electricity trade flows mainly from Laos, which is the subregional net power exporter, to Thailand and Vietnam. The GMS subregional power interconnections are mainly via 110 and 230 kV transmission lines (ADB 2008a). At present, the only high voltage transnational transmission line in the GMS is the

[11] ASCOPE, "The ASEAN Memorandum of Understanding (MoU) on the Trans-ASEAN Gas Pipeline (TAGP)", at: http://www.ascope.org/images/stories/tagp_mou.pdf (8 February 2012).

line from the NT2 hydroelectric power plant in Laos to Thailand, which is the first 500 kV subregional cross-border transmission line. Work is in progress on several such high-voltage transmission lines among other member states.[12]

ASEAN activities for hydropower exploitation go beyond mere infrastructure building and extend into confidence-building measures that facilitate subregional cooperation. For example, the 15th GMS Ministerial Conference that was held in Thailand in 2009 adopted a GMS Road Map for Expanded Cooperation in Energy, which will expand cooperation beyond the electric power sector, improve energy security and promote environmental protection. The roadmap specifies the strategic objectives for expanded GMS energy cooperation and provides a desired policy framework that includes the measures deserving priority. It also presents a concrete and workable work plan that details the specific activities and general timetable for realising these objectives.

Additional initiatives that have been endorsed at the 16th GMS Ministerial Conference in Vietnam include: (i) Core Agriculture Support Program Phase II, which will serve as inputs into the development of the new long-term GMS strategic framework; (ii) Strategic Framework for Connecting GMS Railways, which is a significant step towards an integrated GMS railway system and to promote intermodal transport networks; (iii) Program of Actions for Transport and Trade Facilitation, which is being finalised to facilitate transport and trade in the subregion; and (iv) Strategy and Action Plan for the GMS Southern Economic Corridor, which is meant to provide a vision and framework for developing the Southern Economic Corridor (SEC) in the GMS, improve coordination and mobilise resources.[13]

3.3.4 The Way Forward: Suggestions and Strategies

Further effective action along the lines of the following workable suggestions and strategies may be worth considering for ASEAN countries in order to address challenges ranging from climate change to energy security:

1. Given the complexity of the issue of climate change, it is necessary to focus on particular security issues arising from climate change rather than to deal with them using a multifaceted approach. As mitigation is an important strategy for dealing with climate change, policies that guarantee energy security must be prioritised in the government's agenda.
2. The knowledge base surrounding the impacts of climate change on security needs to be strengthened. The effects of climate change on energy security

[12] ADB, "GMS Development Matrix", at: http://www.adb.org/GMS/Projects/devmatrix.asp?fl=5 (1 October 2011).

[13] ADB, "Greater Mekong Subregion-Progress and Achievements", (August 2010), at: http://www.adb.org/GMS/Program/progress-achievements.asp (1 October 2011).

should be examined in depth, in particular. Governments and academicians should intensify discussions in this regard, and complementary track I and track II exercises will be required accordingly.

3. Regional frameworks need to be established, as sound institutional mechanisms at the national and regional levels would be able to deal with energy security threats effectively. Institutional mechanisms will help equip governments with the capacity to understand and anticipate the threats posed by climate change and to formulate relevant policies and solutions in response. At the regional level, ADB should continue to play its role in regional energy cooperation and economic integration—to provide a comprehensive support package for the preparation of projects, to resolve cross-border agreements and to secure financing for projects. For further energy cooperation and integration in ASEAN and in East Asia, ADB should continue to utilise its expertise and experiences with respect to regional cooperation and integration, and enhance its roles as a facilitator and 'honest broker'.
4. The national government should play a key role in facilitating the establishment of community-based initiatives—an existing tradition in Southeast Asian countries—concerning the management of natural resources and addressing climate change. Since the impacts of climate change threats are local, climate change solutions should originate from the local community, with support from the national government. Grassroots organisations will have rich experience in managing natural resources that should be tapped for this purpose although their success will likely depend on local and national support. A more systematic approach is required to link local efforts to those at higher levels of government. Particular care should be taken to define responsibilities at all government levels and to make these known so that institutional mechanisms across all levels function effectively (Francisco 2008, p. 17).
5. Finally, funds for climate change mitigation should be optimally used. Some funds are already available under the UNFCCC and the Kyoto Protocol for this purpose. However, it is possible that developing countries may be constrained by their lack of knowledge regarding how these are to be obtained despite requirements being posted on relevant websites, such as the United Nations Development Programme (UNDP). The establishment of organisations or institutions with the expertise required to help developing countries develop proposals that take advantage of these funds may be called for (Francisco 2008, p. 17).

3.4 Conclusion

The chapter reviewed the notion of energy security and its measurements and discussed the activities and policies of ASEAN countries with regard to the issues of climate change and energy security. A key aspect of energy security—reliability—was used as a measure of energy diversity to assess the energy security status of ASEAN countries. Data suggested that ASEAN's energy security status was not good due to the region's high dependence on fossil fuels.

ASEAN countries do recognise the various challenges they face due not only to climate change but also energy shortage, and how solutions for one of these challenges may exacerbate targets for the other. Several policies and initiatives are in place to address ASEAN's climate change concerns and improve its energy security. These include the ARF debates on the climate change front and the building of energy infrastructure and the integration of its energy markets for energy security. ASEAN is actively engaged in working out possible solutions for its myriad problems—the recent ARF debates, in Cambodia and Brussels, where ASEAN countries agreed to extend their energy security efforts for combating climate change, are proof of this resolve. Several recommendations, initiatives and policies have been proposed to address these issues, although these are yet to be fully implemented.

References

ADB (Asian Development Bank) (2008a) Energy sector in the greater Mekong subregion. ADB, Manila

ADB (Asian Development Bank) (2008b) GMS: climate makers or climate takers? understanding and responding to the challenges of climate Change, GMS development dialogue, Background paper. ADB, Manila

APERC (Asia Pacific Energy Research Centre) (2004) Electric power grid interconnections in the APEC region. APERC, Tokyo

APERC (Asia Pacific Energy Research Centre) (2007) A quest for energy security in the 21st century: resources and constraints. APERC, Tokyo

Chander S (2000) South Asia growth quadrangle cooperation in the energy sector. Energy Division (West) of the Asian Development Bank, Manila

Chang Y (2009) Global economic crisis and energy security: integrated energy market. In: Hui Ying S, Wai Mun C (eds) Singapore and Asia. World Scientific, Singapore, pp 99–119

Chang Y, Lee JL (2008) Electricity market deregulation and energy security: a study of the UK and Singapore electricity markets. Int J Global Energy Issues 29(1/2):109–132

Chang Y, Yao L (2011) Energy security and energy in a seamless Asia. Panorama: Insights Asian Eur Aff 1:57–71

Chester L (2008) The (default) strategy determining the security of Australia's energy supply. In: Paper for the second international association for energy economics (IAEE) Asian conference: energy security and economic development under environmental constraints in the Asia-Pacific Region. http://www.business.curtin.edu.au/files/Chester_L_2008.pdf. 20 Jan 2012

CIEP (Clingendael International Energy Programme) (2004) Study on energy supply security and geopolitics. Report prepared for DGTREN. CIEP, The Hague

EMA (Energy Market Authority) (2005) Initial findings and views on LNG importation into Singapore, Consultation paper. EMA, Singapore

Francisco H (2008) Adaptation to climate change: needs and opportunities in Southeast Asia. ASEAN Econ Bull 25(1):7–19

Glover D, Lee PO (2008) The environment, climate change and natural resources in Southeast Asia: issues and challenges. ASEAN Econ Bull 25(1):1–6

Gradziuk A (2009) Energy security and climate change: seeking a balance in the new reality. In: Joanna S, Grzegorz S (eds) Energy security and climate change: double challenge for policymakers. Polski Instytut Spraw Międzynarodowych, Warszawa, pp 9–28

Grubb M, Butler L, Twomey P (2006) Diversity and security in UK electricity generation: the influence of low-carbon objectives. Energy Policy 34(18):4050–4062

Hiroshi H (2000) The Mekong: environment and development. United Nations University Press, US

IEA (International Energy Agency) (2007a) Energy security and climate policy: assessing interactions. OECD/IEA, Paris

IEA (International Energy Agency) (2007b) World energy outlook 2007: China and India insights. OECD/IEA, Paris

IEA (International Energy Agency) (2009) World energy outlook. OECD/IEA, Paris

Kleber D (2009) The US Department of Defense: valuing energy security. J Energy Secur (18 June). http://www.ensec.org/index.php?option=com_content&view=article&id=196:the-us-department-of-defense-valuing-energy-security&catid=96:content&Itemid=345. 20 Jan 2012

Runnegar B (1987) Rates and modes of evolution in the Mollusca. In: Campbell KSW, Day MF (eds) Rates of evolution. Allen and Unwin, London, pp 39–60

Sovacool BK, Brown MA (2010) Competing dimensions of energy security: an international perspective. Ann Rev Environ Resour 35:77–108

Sovacool BK, Mukherjee I (2011) Conceptualizing and measuring energy security: a synthesized approach. Energy 36(8):5343–5355

Sovacool BK, Mukherjee I, Drupady IM, D'Agostino AL (2011) Evaluating energy security performance from 1990 to 2010 for eighteen countries. Energy 36(10):5846–5853

Stirling A (1994) Diversity and ignorance in electricity supply investment: addressing the solution rather than the problem. Energy Policy 22(3):195–216

Stirling A (2010) Multicriteria diversity analysis: a novel heuristic framework for appraising energy portfolios. Energy Policy 38(4):1622–1634

WWF (World Wide Fund for Nature) (2009) The greater Mekong and climate change: biodiversity, ecosystem services and development at risk. WWF, Bangkok

Yergin D (1988) Energy security in the 1990s. Foreign Aff 67(1):110–132

Abbreviations

ADB	Asian Development Bank
AIMS	ASEAN Interconnection Master Plan Study
APERC	Asia Pacific Energy Research Centre
APG	ASEAN Power Grid
APGCC	ASEAN Power Grid Consultative Committee
ARF	ASEAN Regional Forum
ASCOPE	ASEAN Council on Petroleum
ASEAN	Association of Southeast Asian Nations
CIEP	Clingendael International Energy Programme
CMP	Conference of the Parties Serving as the Meeting of the Parties
CPEC	Cambodia Petroleum Exploration Co
EMA	Singapore Energy Market Authority
EU	European Union
GHG	Greenhouse gas
GMS	Greater Mekong Subregion
HAPUA	Heads of ASEAN Power Utilities/Authorities
HSS	School of Humanities and Social Sciences
IAEE	International Association for Energy Economics

IEA	International Energy Agency
kV	Kilovolts
LNG	Liquefied natural gas
mmbd	Million barrels per day
MOU	Memorandum of understanding
mtoe	Million tonnes of oil equivalent
MW	Megawatts
NEA	National Environment Agency
NT2	Nam Theun 2 hydroelectric power project
NTS	Non-traditional security
NTU	Nanyang Technological University
OECD	Organisation for Economic Co-operation and Development
RSIS	S. Rajaratnam School of International Studies
SEC	Southern Economic Corridor
TAGP	Trans ASEAN Gas Pipeline
UK	United Kingdom
UNDP	United Nations Development Programme
UNFCCC	United Nations Framework Convention on Climate Change
WWF	World Wide Fund for Nature

Author Biographies

Youngho Chang (Singapore): Assistant professor of Economics at the Division of Economics, Nanyang Technological University (NTU), Singapore. Since 1999, he has taught resource and energy economics, environmental economics and macroeconomics at the National University of Singapore and NTU. He has also contributed to various academic journals, including *Econometric Theory*, *Energy Policy*, *Economics Letters*, *International Journal of Global Energy Issues* and *International Journal of Electronic Business Management*. He received a BASc in landscape architecture from Seoul National University, South Korea; an MA in economics from Yonsei University, South Korea; and a PhD, also in economics, from the University of Hawaii at Manoa, US.

Lixia Yao (Singapore): PhD candidate at the S. Rajaratnam School of International Studies (RSIS), Nanyang Technological University (NTU), Singapore. She has a Bachelor's degree from Dalian Maritime University, China, and Master's degree from Vrije Universiteit Brussel, Belgium. She was also a lecturer at Hebei Professional College of Political Science and Law, China. Her current research focuses on China's energy security and energy policies. Publications include: "Energy Security and Energy in a Seamless Asia", in: *Panorama: Insights into Asian and European Affairs*, 1 (2011); "Energy and the Greater Mekong Sub-region (GMS): Cooperation, Competition and Development", in: *Dealing with Energy Vulnerabilities: Case Studies of Cooperation and Collaboration in East Asia* (forthcoming in 2012).

Chapter 4
The Socioeconomic Impact of Energy Security in Southeast Asia

Maria Nimfa F. Mendoza

Abstract This chapter aims to analyse some of the energy security issues facing ASEAN (Association of Southeast Asian Nations) countries and their socioeconomic impacts, in the context of developments in the global energy markets. ASEAN countries subscribe to the core strategies of the development of new sources, diversification of supply and greater use of renewables; promotion of competitive markets; greater energy efficiency in use; and regional cooperation, trade in energy goods and services, and cross-country investments. Key challenges are the reduction of fossil fuel subsidies and improving access to modern energy services. The adoption of renewable energy in the region is hampered by its high cost.

Keywords ASEAN · Biofuels · Electricity · Energy security · Energy self-sufficiency · Fossil fuel subsidies · Modern energy services · Regional cooperation · Renewable energy · Socioeconomic impacts

M. N. F. Mendoza (✉)
School of Economics, University of the Philippines (Diliman), Quezon City, The Philippines
e-mail: ma_nimfa.mendoza@up.edu.ph, mnfm196@yahoo.com
www.econ.upd.edu.ph

M. Caballero-Anthony et al. (eds.), *Energy and Non-Traditional Security (NTS) in Asia*, SpringerBriefs in Environment Security, Development and Peace 1,
DOI: 10.1007/978-3-642-29706-9_4,

4.1 Introduction

There is close interconnection between energy security,[1] socioeconomic development and environmental protection. Since energy is an important input to economic activities, reliable, adequate and affordable energy supplies are essential to support the economic growth and competitiveness of fast-growing Southeast Asian countries. Concerns about global climate change and concomitant environmental damage, such as increased air pollution from more transport vehicles and greater power generation, have prompted a push in the region towards greater use of renewable energy and low-carbon technologies. As a net importer of oil, energy security for the region includes efforts toward diversification of energy sources and policy responses to exogenous shocks and disruptions, such as a fall in demand due to a global financial crisis or a supply disruption due to political problems in the Middle East.[2] As most Southeast Asian countries are either low- or middle-income countries, access to energy by the poor is an additional dimension of energy security in the region.

Energy being a key requisite of economic activities, such as for the production of goods and services as well as household consumption, the welfare of energy users is linked to developments in the energy sector. The socioeconomic effects of changes in energy prices affect both energy producers and consumers. The ultimate socioeconomic impact of such changes is on the consumption of goods and services by households, including those who derive income from energy-producing sectors, by way of effects on access to energy, household incomes, and the relative prices of goods and services.

The chapter will focus on the socioeconomic impacts of recent developments in the global energy market roughly during the 2008–2011 period in Southeast Asia. The discussion will be in the context of rising energy prices, which is generally being perceived as a permanent trend or shift from the low fuel prices that were seen in the past two decades. Recognising the limited global reserves of fossil fuels and their high dependence on imported oil, the volatility of world prices of energy especially oil, the growing concerns about the local environment and global climate change, and the increasing demand for energy due to increased economic growth, Southeast Asian countries have deemed energy security as a major regional concern (ASEAN 2007). Major policy thrusts can be divided into three general areas: (i) the development and greater use of cleaner and lower-emission

---

[1] Energy security has been defined in terms of the availability, reliability, affordability and sustainability of energy supplies. Sustainability is considered in the context of "low emissions of greenhouse gas and other pollutants; minimal contribution to local, regional or global threats to environmental quality; and protection of energy systems from impacts of a changing climate" (Elkind 2010, p. 122).

[2] High oil prices due to supply disruptions or other exogenous supply shocks may necessitate different policy responses. For example, protection against piracy and terrorism in the Malacca Straits, which is a major shipping lane in Southeast Asia, is enhanced by joint surveillance activities of Singapore, Indonesia and Malaysia.

technologies, including the use of renewable energy and deliberations on the nuclear option; (ii) promotion of efficiency in energy use; and (iii) the pursuit of more open trade, facilitation and regional cooperation in the energy sector. Another social concern of Southeast Asian countries, particularly for lower-income countries, is access to modern energy services, such as electricity and clean cooking facilities.

The country experiences of the ten ASEAN (Association of Southeast Asian Nations) member countries—Brunei, Cambodia, Indonesia, Laos, Malaysia, Myanmar, The Philippines, Singapore, Thailand and Vietnam—will be used to illustrate socioeconomic impacts in the region. In 2009, these countries had a total population of 592 million, with a per capita GDP of US$2,533 (in current US$).[3] These countries differ with regard to initial conditions, such as in terms of resources, general income levels, stages of economic development and energy policies, and therefore it is expected that differential impacts would be seen among them, at least where the magnitudes of their responses are concerned.

The chapter looks at the energy security concerns of Southeast Asian countries and how they address these concerns, in particular the socioeconomic aspect of energy security. The chapter aims to assess the current state of energy markets in Southeast Asia, and analyse the energy policies and energy security issues facing ASEAN countries and their response to developments in the global energy markets, with special emphasis on the socioeconomic effects of these responses. The chapter is divided into seven sections. Section 4.2 gives an overview of the economy and energy sector in Southeast Asia. Section 4.3 discusses the fossil fuel subsidy policies of various Southeast Asian countries. These sections lay out the context for a discussion of the socioeconomic impacts of recent developments in the global energy market in the subsequent two sections—Sect. 4.4 discusses the macroeconomic effects of these developments while Sect. 4.5 analyses their microeconomic effects. Section 4.6 discusses policy options and questions for the future and Sect. 4.7 concludes the chapter by offering an overview of discussion and options.

4.2 Energy Markets in ASEAN Countries

4.2.1 Energy Options

Southeast Asia is rich in resources such as natural gas, coal and hydropower. Brunei is a major oil and gas producer and exporter while Indonesia and Malaysia have abundant oil and natural gas reserves. Malaysia is a net oil-exporting country, but Indonesia, which is Southeast Asia's biggest oil user, is a net importer of oil

[3] ASEAN, "Selected Basic ASEAN Indicators: as of 15 February 2011", in: *ASEAN Statistics* (2009), at: http://www.aseansec.org/19226.htm (7 January 2012).

since 2004. Indonesia, with its large population, is Asia's top gasoline and diesel importer. Cambodia, the Philippines, Singapore and Thailand are net energy importers while Brunei, Indonesia, Malaysia, Myanmar and Vietnam are net energy exporters.[4] In 2006, significant oil and natural gas reserves were discovered in Myanmar.[5] Exploration efforts for oil and natural gas are similarly being undertaken in Vietnam, the Philippines and the disputed Spratly Islands. Although it does not have its own oil reserves, Singapore is a major refining centre, with most of its output being exported, and a major petrochemical producer. Indonesia, and, to a lesser extent, the Philippines produce and export coal. The Philippines is also the second largest producer of geothermal energy, second only to the US. Meanwhile, Indonesia, with larger geothermal energy potential, has ambitious plans to generate 3,997 MW of geothermal power by 2014 (Tay and Fang 2011).

Notwithstanding the abundance of its natural resources, the Southeast Asian region remains heavily dependent on crude oil imports from the Middle East, with most countries relying on oil and gas as main sources of primary energy. The share of fossil fuels and renewable sources in total primary energy supply as of 2005 is given in Fig. 4.1. The lower-income countries tend to use more biomass, which has traditionally been used for cooking and is often viewed as a non-commercial rural energy source or as the poor man's fuel. In 2005, for instance, biomass met nearly half of the primary energy needs in Cambodia, Myanmar and Vietnam. A more balanced mix of energy sources is seen in Indonesia and the Philippines, with the latter using a greater proportion of hydropower and geothermal energy relative to other ASEAN countries. Figure 4.2 depicts the primary energy supply of ASEAN countries in 2007, providing a glimpse of the share of fossil fuels and renewable sources. It should be noted that the high proportion of renewable resource use seen for Cambodia, Vietnam and Myanmar is not indicative of a shift from fossil fuels to renewables but a reflection of the use of biomass and a lack of access to modern energy services. Indeed, such reliance on the use of biomass for cooking not only indicates energy poverty at the household level but also raises environmental concerns regarding the health effects of its use in inefficient stoves due to subsequent indoor air pollution. Possible alternative cooking options could be liquefied petroleum gas (LPG) stoves, biogas systems and advanced biomass cook stoves.

Most Southeast Asian countries—Vietnam, Indonesia, the Philippines, Malaysia, Thailand and Myanmar—are considering the nuclear option although lead times for starting operations (i.e. setting up the legal and regulatory infrastructure as well as undertaking the actual construction) of a nuclear plant in countries where nuclear technology is not yet established typically range from 10 to15 years. Vietnam, likely to be the first Southeast Asian country to have a nuclear power plant, plans to have four nuclear generation plants with a total capacity of

[4] For statistics on net energy imports as a percentage of energy use, see ADB 2011, p. 259. For Laos, data was not available.

[5] Nguyen, Anuchit, "Gas and Oil Discovered in Myanmar", in: *International Herald Tribune* (6 August 2006), at: http://www.iht.com/articles/2006/08/06/Bloomberg/sxptt.php (22 August 2008).

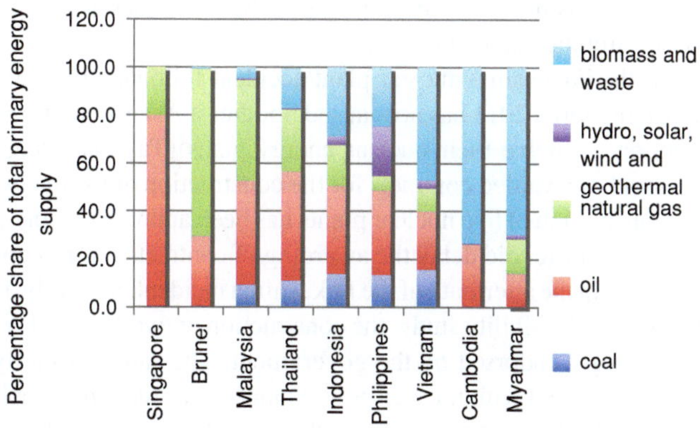

Fig. 4.1 Share of total primary energy supply by source, 2005 (For Laos, data were not available). *Source* UNDP 2007, pp. 306–309. *Note* Fossil fuels include coal and coal products, crude, natural gas liquids, feedstocks, petroleum products and natural gas (Singapore, as a small island nation, has no hydropower or geothermal resources and thus uses mostly fossil fuels). Renewable resources include hydropower, geothermal power, combustible renewables, waste, and solar and wind power, but exclude nuclear energy. (The high proportion of renewable energy resources for Vietnam, Cambodia and Myanmar are due to high use of biomass as cooking fuel.)

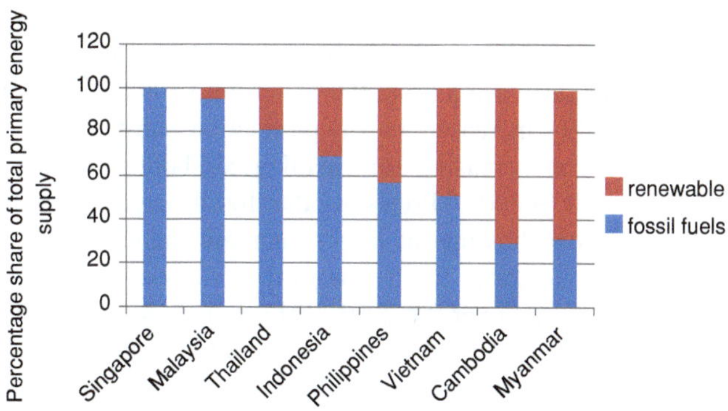

Fig. 4.2 Share of total primary energy supply, fossil fuels and resources, 2007 (For Brunei and Laos, data were not available). *Source* UNDP 2010, pp. 168–171

8,000 MW in operation by 2025.[6] Similarly, Russia's Federal Agency on Atomic Energy (now Rosatom) signed a deal in 2007 with the Burmese government to build Myanmar's first nuclear research centre with a 10 MW light water

[6] Huyen, Ngoc, "Vietnam to Build Two Nuclear Power Plants", in: *VietNamNet Bridge* (14 April 2008), at: http://vietnamnet.vn/service/printversion.vnn?article_id=1054251 (7 January 2012).

reactor.[7] Nuclear feasibility studies are also being undertaken in Indonesia, Thailand, the Philippines and Malaysia.

Interestingly, the Fukushima nuclear plant accident following the earthquake and tsunami in Japan in March 2011 has not caused Southeast Asian countries considering the nuclear option to shelve their nuclear energy plans (Tay and Fang 2011). For instance, Vietnam has awarded contracts for the construction of its first two plants. In Thailand, the construction of five nuclear plants has been approved by the government although its prime minister did delay the awarding of contracts for the construction of its first nuclear plant in the aftermath of the Fukushima incident. Similarly, in Malaysia, pending the result of a feasibility study, the construction of two 1,000 MW reactors by 2022 is being strongly endorsed by the government. The safety, security and environmental risks associated with nuclear power plants have remained unchanged notwithstanding the Fukushima accident. Nevertheless, it has raised public awareness of the risks associated with nuclear energy and made governments increasingly cognisant that sufficient safeguard measures would have to be put in place during the construction, operations, waste disposal and decommissioning of such power plants should they decide to pursue their plans for nuclear energy.

4.2.2 Economy and Energy

There were two oil shocks in the 1970s—the Arab oil embargo in 1973 and another oil price spike during 1979–1981 due to the Iranian revolution and the Iran-Iraq war. Oil prices maintained a decreasing trend thereafter until 1998, with a sharp drop in 1986 when Saudi Arabia increased its oil production substantially. From 1999 onwards, there has been an increasing trend in oil prices, both nominal and real, until the global financial crisis in 2008.[8] Prices of other fossil-based energy have risen in recent years too, with coal showing the largest increase and natural gas less so than spot crude. Figure 4.3 illustrates the trends in the prices of energy—particularly crude, natural gas and coal—over the 2001–2011 period. As seen from the graph, although prices reached a peak in 2008, at the start of the global financial crisis that subsequently led to a fall in energy prices, by mid-2011, prices had recovered to nearly pre-crisis levels. It is widely believed that high energy prices are likely to persist and that the principal drivers of this trend are fundamental factors of supply

[7] "Russia to Build Nuclear Reactor in Myanmar", in: *NewScientist* (15 May 2007), at: http://www.newscientist.com/article/dn11856-russia-to-build-nuclear-reactor-in-myanmar.html (7 January 2012); "Russia to Build Nuclear Reactor in Myanmar", in: *Reuters* (15 May 2007), at: http://www.reuters.com/article/2007/05/15/us-russia-myanmar-nuclear-idUSL1565024820070
515 (7 January 2012).

[8] "Oil Prices", in: *OECD Factbook 2010: Economic, Environmental and Social Statistics*, at: http://www.oecd-ilibrary.org/sites/factbook-2010-en/05/02/03/index.html?contentType=&itemId=/content/chapter/factbook-2010-44-en&containerItemId=/content/serial/18147364&accessItemIds=&mimeType=text/html (7 November 2011).

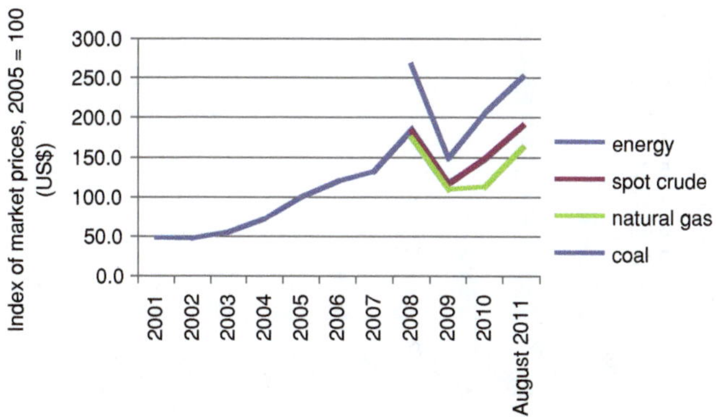

Fig. 4.3 Indices of market prices for energy commodities, 2001–2011. *Source* IMF 2011, (IMF, "Table 1a. Indices of primary commodity prices, 2001-August 2011", at: http://www.imf.org/external/np/res/commod/Table1.pdf (6 September 2011); IMF, "Table 2. Indices of market prices for non-fuel and fuel commodities, 2008-August 2011", at: http://www.imf.org/external/np/res/commod/Table2.pdf (6 September 2011)

and demand. The increase in demand, due to an expansion of the world economy, has been greater than that in supply and this imbalance is likely to persist in the next few years. Demand from emerging market economies, such as China and India, is expected to remain robust and global supply of oil relatively tight. Other factors—the weak US$, which is causing investors to flee to commodities, and geopolitical tensions involving oil-producing countries resulting in fears of production disruptions—are also playing a contributory role.

Energy is mainly used as fuel for transport and power generation. As expected, within the Southeast Asian region, high-income countries consume more energy per capita than lower-income countries (Fig. 4.4). The high per capita energy use in Brunei can be explained by its high energy subsidies while the relatively low per capita energy use in the Philippines can be explained by its high electricity prices.

There is wide variation in the economic development, country population and per capita income levels among the Southeast Asian countries (Figs. 4.5 and 4.6). For instance, at one end are the richer but smaller ASEAN countries—Singapore (2009 population: 4.988 million), which is considered a developed country, and Brunei (2009 population: 406,000), with its vast oil reserves—while at the other end are less developed countries, such as Laos, Cambodia and Myanmar. Average GDP per capita of ASEAN in 2009 was US$2,533 (US$4,873 purchasing power parity [PPP]-adjusted).[9] Thailand, Indonesia, the Philippines, Vietnam and Myanmar have relatively large populations of over 50 million—Indonesia, with a population of 232 million in 2009, accounts for about 40 % of the ASEAN population.[10] Figure 4.7

9 ASEAN, "Selected Basic ASEAN Indicators: as of 15 February 2011", in: *ASEAN Statistics* (2009), at: http://www.aseansec.org/19226.htm (7 January 2012).

10 Ibid.

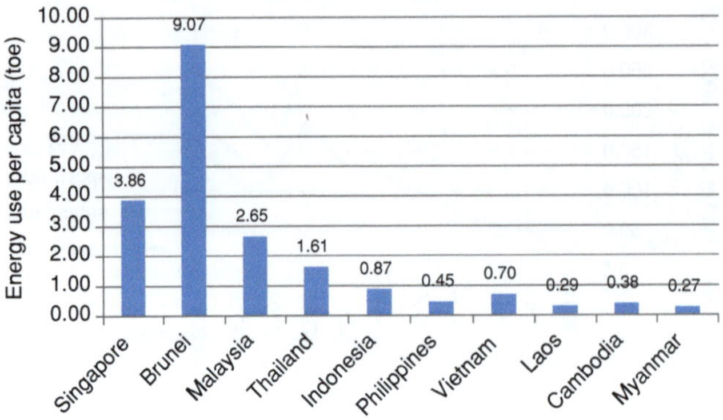

Fig. 4.4 Energy use per capita, toe, 2008 (For Laos, the value is for 2006). *Source* ADB (Asian Development Bank) 2011, p. 134, 260. *toe* tons of oil equivalent

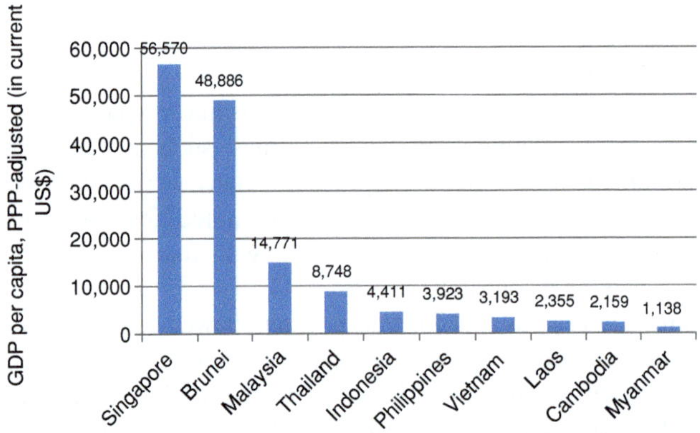

Fig. 4.5 GDP per capita, PPP-adjusted (in current US$), 2010. *Source* ADB 2011 (For GDP data on all countries other than Myanmar): 162; ASEAN 2009 (For data on Myanmar, see ASEAN, "Selected Basic ASEAN Indicators: as of 15 February 2011", in: *ASEAN Statistics* (2009), at: http://www.aseansec.org/19226.htm (7 January 2012)). *GDP* gross domestic product, *PPP* purchasing power parity

illustrates the generally high economic growth rates of ASEAN countries over the 1990–2000 and 2000–2009 periods vis-à-vis the performance of low-income, middle-income and high-income countries in the world.

The GDP per unit of energy use in 2008 for ASEAN countries is shown in Fig. 4.8. The GDP per unit of energy use, which reflects the structure and energy intensiveness of the economy, is an indicator of energy productivity.[11] Singapore

[11] The inverse of GDP per unit of energy, which is the amount of energy used to produce a unit of GDP, indicates the energy intensity of production.

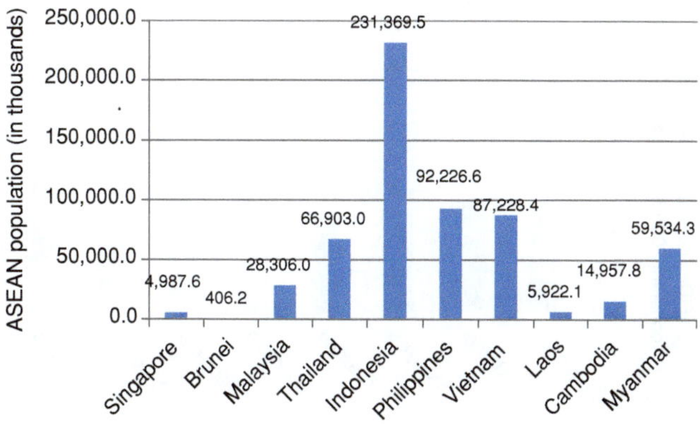

Fig. 4.6 Population of ASEAN countries. *Source* ASEAN 2009 (ASEAN, "Selected Basic ASEAN Indicators: as of 15 February 2011", in: *ASEAN Statistics* (2009), at: http://www.aseansec.org/19226.htm (7 January 2012)). *ASEAN* Association of Southeast Asian Nations

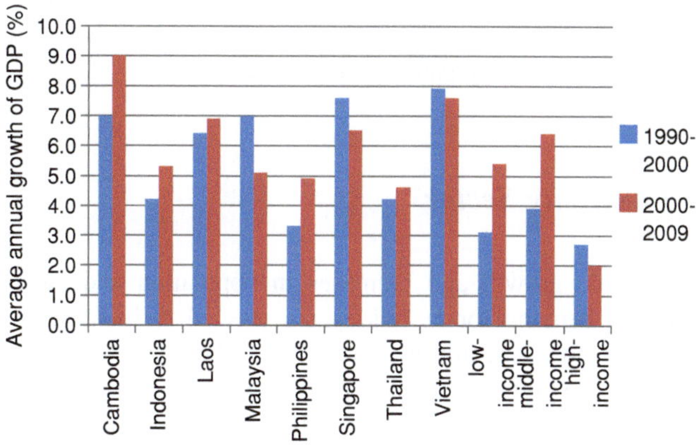

Fig. 4.7 Average annual growth rate of real GDP, 1990–2000 and 2000–2009 (For Brunei and Myanmar, data were not available). *Source* World Bank 2011, pp. 194–196. *Note* Graphs for the low-, middle- and high-income countries cover those for the world and are not limited to ASEAN countries. *GDP* gross domestic product

has the highest GDP per unit of energy use among ASEAN countries, reflecting a less energy-intensive economy structure where the service-oriented industry dominates. The relatively high GDP per unit of energy use in the Philippines, with its deregulated oil and electricity sectors, may reflect the response of its production sector to its high or subsidy-free electricity and fuel prices that induce lesser use of energy. In economies where the agricultural sector or service sector is relatively large and the industrial sector small, the energy intensity of production tends to be low. In this case, the low energy intensity is consequently reflected in a higher energy productivity (high GDP/energy ratio) indicator, which is not necessarily

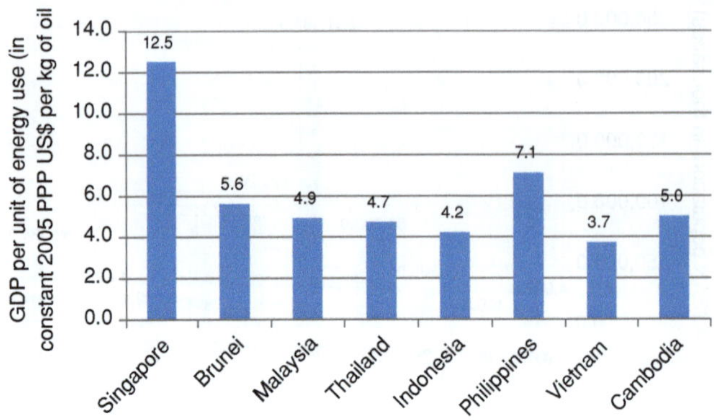

Fig. 4.8 GDP per unit of energy use, in constant 2005 PPP US$ per kilogram of oil equivalent, 2008. (For Laos and Myanmar, data were not available; for Brunei, the value is for 2007.) *Source* ADB 2011, p. 257. *GDP* gross domestic product, *PPP* purchasing power parity

due to energy efficiency but more due to the structure of the economy. For this reason, Cambodia's GDP/energy ratio may reflect more the structure of its economy, as it has a relatively small industrial sector.

4.2.3 The Electricity Sector

With high economic growth and increases in population and electrification in ASEAN countries, demand for electricity is expected to increase in the region. Investments in the power sector would therefore be required to support future economic growth. The higher fossil fuel prices, together with government policies that provide incentives for renewables as well as global climate change mitigation concerns, are signalling a shift in the market towards greater investments in electricity generation using renewable sources and other low-carbon technologies.

Figure 4.9, which presents the sources of electricity in the ASEAN at a regional level, reveals that nearly half of the region's electricity requirement is sourced from natural gas. Coal accounts for just over one-fourth of the source mix and about 15 % of electricity comes from renewables in the region. The sources of electricity, by percentage share, for the Southeast Asian countries as of 2008, are depicted in Fig. 4.10. Natural gas emerges once again as the major source of electricity—accounting for at least 32 % of electricity production in the Philippines, at the lower end, and nearly 99 % in Brunei, at the higher end—for most countries except for Indonesia, Cambodia and Myanmar.

The choice of source of electricity partly reflects the resource endowments of the countries: natural gas for Brunei, coal for Indonesia, and hydropower for Vietnam and Myanmar. For similar reasons, a significant proportion of 'other

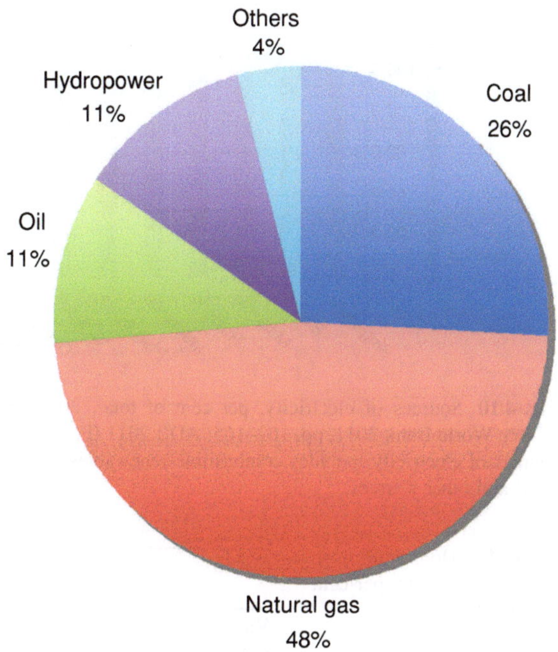

Fig. 4.9 Sources of electricity for the ASEAN region, 2008 (For Laos, data was not available). *Source* World Bank 2011, pp. 162–165; ADB 2011 (For data on Brunei), p. 249. *Note* 'Others' in sources of electricity includes combustible renewables and wastes as well as geothermal, solar, wind and other sources

sources' for the Philippines is geothermal energy. Currently, no ASEAN country sources electricity from nuclear power.

Household electrification rates in ASEAN countries, as shown in Table 4.1, suggest that providing access to electricity services remains a major social concern for most member countries not belonging to the higher-income group (Singapore, Brunei, Malaysia and Thailand). This is particularly true for lower-income countries, such as Myanmar, Laos and Cambodia. Data on per capita annual electricity consumption in Southeast Asian countries (Fig. 4.11) suggest that consumption is strongly associated with income—large disparities are apparent between the richer and poorer countries in the region.

The market structure of the electricity industries in Southeast Asian countries, with the exception of those in Singapore and the Philippines, conform more closely to the old model of monopoly franchise with a high degree of vertical integration, full requirements service (wherein electricity power is provided in a bundle by the local utility company), and rate regulation to ensure reasonable cost recovery and returns on capital investment. In Singapore and the Philippines, the electricity markets have been undergoing restructuring, with Singapore further along in the process—Singapore is currently in the retail liberalisation phase[12] while the Philippines, where the restructuring process began in 2001, expects to

[12] "Electricity Surges Ahead", in: *Singapore Notes* (1 July 2011), at: http://singaporedesk.blogspot.com/2011/07/electricity-surges-ahead.html (7 January 2012).

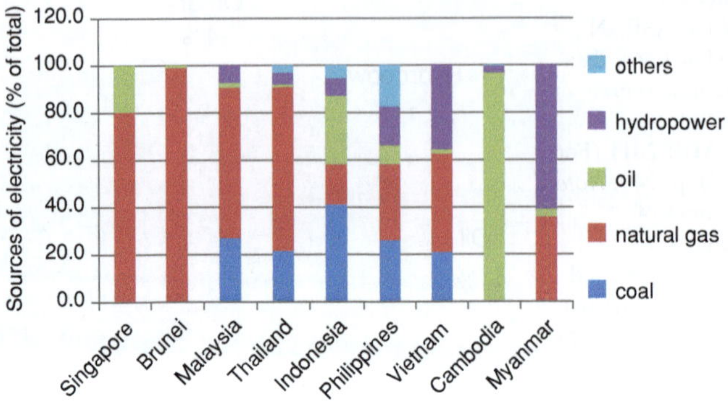

Fig. 4.10 Sources of electricity, per cent of total, 2008 (For Laos, data was not available). *Source* World Bank 2011, pp. 162–165; ADB 2011 (For data on Brunei), p. 249. *Note* 'Others' in sources of electricity includes combustible renewables and wastes as well as geothermal, solar, wind and other sources

Table 4.1 Household electrification rate, per cent of households

Country	Year	Household electrification rate (% of households)
Vietnam	2005	96.1
Indonesia	2007	91.1
The Philippines	2008	83.3
Myanmar	2002	47.0
Laos	2002	46.3
Cambodia	2005	20.5

Source ADB 2011, p. 250

start with the open access and retail competition phase in the third quarter of 2012.[13]

The rationale for most restructurings in the electricity sector worldwide, as also for ASEAN countries, has been a weakening of the natural monopoly argument based on economies of scale (Hogan 2009). A further major impetus for restructuring in the case of the Philippines was the substantial losses being incurred by the state-owned National Power Corporation (NPC) that put into question its long-term viability without huge government support. Some of the key features of such a shift to a more competitive structure would include the unbundling of generation, transmission, distribution and retail supply; elimination of cross-subsidies across customer classes; competition in generation and through the wholesale spot electricity market; and regulation of transmission services that retain features of a natural monopoly partly to ensure open access and system reliability.

[13] Remo, Amy R., "New Timetable for Power Program Set", in: *Philippine Daily Inquirer* (1 November 2011): B1.

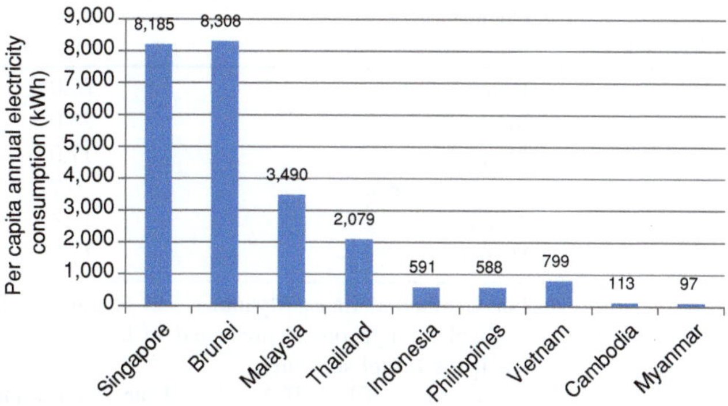

Fig. 4.11 Per capita annual electricity consumption, kWh, 2008 (For Laos, data was not available). *Source* ADB 2011, p. 250. *kWh* kilowatts hour

4.3 Fossil Fuel Subsidies in ASEAN Countries

This section presents a survey of energy-related government policies, with a focus on fossil fuel subsidies. The fossil fuel consumption subsidy rates of ASEAN countries are given in Table 4.2. Except for Singapore and, to a lesser extent, the Philippines, the governments of Southeast Asian countries provide fuel subsidies. Subsidy rates are high (range, 17.0–26.0 %) in oil-exporting countries, such as Brunei, Indonesia and Malaysia. In contrast, in Singapore and the Philippines, where the oil sector (both upstream and downstream) and the electric power industry are deregulated, pump gas prices and electric power rates hardly reflect any subsidy.

In Southeast Asia, energy subsidies, when they do exist, are largely for fuel and electricity. Fossil fuel subsidies in ASEAN countries are generally directed at gasoline, diesel and socially sensitive products, such as LPG and kerosene. Fuel subsidies in developing countries are generally skewed towards diesel since diesel is mostly used by the public or commercial transport sector—jeepneys in the Philippines, buses, truckers, fishing boats, ships and the like. The cheapest fuel per gallon in many Asian countries is kerosene, which is used for lighting and cooking by poorer households. LPG is widely used for cooking in some Southeast Asian countries instead of the more expensive electricity. LPG and compressed natural gas (CNG) are also being promoted as alternative fuels for transport purposes, with LPG enjoying greater and wider public acceptance although it still accounts for only a small portion of the market. Since increases in diesel, kerosene and LPG prices tend to have a greater impact on the consumption of poorer households, these prices are usually subject to vigilance by civil society groups in these countries. Notably, these fuels are fossil based and their use is driven more by their lower prices than environmental considerations. Some ASEAN countries, including Indonesia, Malaysia, Thailand and Vietnam, also subsidise electricity (IEA 2009, pp. 542–543).

Table 4.2 Fossil fuel consumption subsidy rates as a proportion of the full cost of supply, 2009

Country	Average subsidisation rate (%)
Brunei	26.0
Indonesia	25.0
Malaysia	17.0
The Philippines	0.7
Thailand	12.6
Vietnam	11.3

For Cambodia, Laos, Myanmar and Singapore, data were not available. Singapore is presumed to have nil or at most very small fossil fuel subsidies
Source IEA 2010, (IEA, "Fossil-fuel Consumption Subsidy Rates as a Proportion of the Full Cost of Supply, 2009", in: *World Energy Outlook 2010* (2010), at: http://www.iea.org/subsidy/index.html (29 September 2011) at: http://www.iea.org/subsidy/index.html)

Though direct fossil fuel subsidies are nil or small in Singapore and the Philippines, indirect fossil fuel subsidies may implicitly exist in both countries in the form of electricity subsidies for lower-income households. The Singaporean government through its public housing authority, the Housing and Development Board (HDB), provides the so-called 'Utilities-Save (U-Save)' Rebates, which is part of a 5-year government service tax (GST) offset package that was announced in 2007 with final payouts in January 2012.[14] U-Save rebates are credited directly to the utility account of the premises, with unused rebates being rolled over to the following months. About 800,000 HDB Singaporean households benefit from these rebates. Households living in smaller HDB flats[15] receive additional rebates, which are not directly tied to electricity consumption. In Singapore then, rebate schemes are geared towards helping lower- and middle-income households. The U-Save Rebate scheme's design, with its rollover feature for unused rebates, is efficient in the sense that it is non-distortionary—households still face the same marginal price for electricity, which is the going market price; the rebates work effectively as a direct income subsidy.[16]

[14] The Singapore Government, "'Grow and Share' Package Overview", at: http://www.growandshare.gov.sg/Overview.htm (4 November 2011); The Singapore Government, "Utilities-Save (U-Save) Rebate", at: http://www.growandshare.gov.sg/FAQs.htm#5 (4 November 2011); and "800k Households Get $35 m in U-Save Rebates", in: *The Straits Times* (30 June 2011), at: http://www.straitstimes.com/BreakingNews/Singapore/Story/STIStory_685753.html (7 January 2012).

[15] The rebate amount depends on the number of rooms in the flat.

[16] Support was earlier provided as Utility Schemes in the early 2000s. The increase in demand for electricity due to the U-Save Rebate scheme is due only to the income effect, there being no relative price change due to the scheme. As microeconomic theory has shown, an income subsidy is superior to a subsidy-in-kind.

In the Philippines, the government undertakes 'missionary electrification' in remote and off-grid areas that are not connected to the main transmission grid via the National Power Corporation's Small Power Utilities Group (NPC-SPUG). Most operational expenses of the SPUG, which oversees the programme, are for fuel supply to mostly diesel-fired power plants.[17] In addition, the Philippine government also implements a compensating mechanism for targeted poorer households—mainly for the purposes of political expediency and the public perception that the government is doing something for the poor—by providing lifeline rates to households consuming below a threshold level (low users) of grid-connected electric power. However, such electricity lifeline rates, financed by cross-subsidies within distribution utility areas, are more likely to benefit middle-class families in urban areas. The financing scheme has the disadvantage that poorer distribution utility areas would have less capacity to raise electricity tariffs to pay for the lifeline rate scheme.[18] Studies show that lifeline rates have significant leakages especially if the threshold levels are set high (with monthly consumption above 50 kWh)—the higher the threshold level, the higher the cost of the lifeline rate programme, the greater the proportion going to the non-poor, and consequently the greater the distortion of incentives meant to improve energy efficiency. Moreover, nearly 15 % of the total households in the Philippines are likely to be among the poorest of the poor in remote rural areas with no access to electricity and therefore do not enjoy the benefits of lifeline rates. In contrast to Singapore's U-Save Rebate scheme, the lifeline rates for grid-connected electricity consumers and subsidised rates provided in the SPUG areas are distortionary—electricity consumers face a marginal price lower than what would have been applicable without these intervention schemes—and would generally induce beneficiaries to consume more electricity than otherwise.[19]

Fuel subsidies lead to higher demand and, if available to the general population, can impose fiscal difficulties on government budgets when oil prices shoot up, as can be seen from the experiences of most Southeast Asian countries. With the increase in international oil prices to record highs in 2008 amid fixed domestic fuel prices, the cost of subsidies to governments increased—in some cases to levels where the subsidy cost could not be sustainably financed by the government given the rising fuel prices. For instance, fuel prices in Indonesia and Malaysia were increased in 2008 primarily due to the fiscal burden of fuel subsidies on the

[17] Remo, Amy R., "Napocor Needs P15B for Off-grid Plants", in: *Philippine Daily Inquirer* (31 October 2011): B1.

[18] Ideally, the lifeline rate scheme, as a redistributive mechanism, should be financed by the national government or through a universal charge paid by consumers of grid-connected electricity.

[19] In the Philippines, the lifeline rates and subsidised tariff rates in the SPUG areas would have both relative price effects (substitution effects, where electricity substitutes other goods as it is relatively cheaper) and income effects (the subsidised lower electricity tariffs free up income, increasing purchasing power, so that consumers can buy more goods and services, including more electricity).

government budget. Despite the price adjustments, however, fuel prices at the retail level mostly remained below market rates. The experiences of eight Southeast Asian countries, excluding Singapore and the Philippines, with fossil fuel subsidies during the 2008–2011 period are discussed next.

4.3.1 Indonesia

The Indonesian government was estimated to have spent about US$20 billion (180.31 trillion rupiahs) or about 4 % of GDP on fuel subsidies in 2008[20] and nearly US$6.5 billion (60.4 trillion rupiahs) on electricity subsidy in 2009.[21] These estimates are for subsidies after the price of fuel was adjusted in May 2008, when subsidised fuel prices increased on average by 28.7 % (prices rose by 33 % for premium fuel, 28 % for diesel and 25 % for kerosene). To help the poor cope with the impact of such fuel price hikes, the government has put in place compensation programmes, including direct cash assistance, for 19.1 million poor families. However, recognising the inefficiency of fossil fuel subsidies, Indonesia has embarked on fossil fuel subsidy reform albeit gradually so as to cushion the poor from the effects of such reform. In 2010 and 2011, fuel subsidy was estimated to have been 8 % of public expenditure (government budget) (Fiscal Policy Office 2011, p. 14).[22]

---

[20] Alfian; Suharmoko, Aditya, "Govt Proposes to Reduced Fuel Subsidy", in: *The Jakarta Post* (16 August 2008), at: http://www.thejakartapost.com/news/2008/08/16/govt-proposes-reduced-fuel-subsidy.html (6 February 2012); Bradsher, Keith, "Fuel Subsidies Overseas Take a Toll on U.S.", in: *The New York Times* (28 July 2008), at: http://www.nytimes.com/2008/07/28/business/worldbusiness/28subsidy.html (7 January 2012).

[21] Alfian; Suharmoko, Aditya, "Govt Proposes to Reduced Fuel Subsidy", in: *The Jakarta Post* (16 August 2008), at: http://www.thejakartapost.com/news/2008/08/16/govt-proposes-reduced-fuel-subsidy.html (6 February 2012).

[22] For more information, see Bradsher, Keith, "Fuel Subsidies Overseas Take a Toll on U.S.", in: *The New York Times* (28 July 2008), at: http://www.nytimes.com/2008/07/28/business/worldbusiness/28subsidy.html (7 January 2012); Bradsher, Keith, "Prices of Food and Gas Take a Toll in Asia", in: *The New York Times* (23 July 2008), at: http://www.nytimes.com/2008/07/23/business/worldbusiness/23inflate.html (7 January 2012); "Indonesia Sees Costly Oil Stretching Fuel Subsidy", in: *Reuters* (20 April 2008), at: http://www.reuters.com/article/2008/04/20/us-ief-indonesia-idUSL2010104120080420 (6 February 2012); Alfian; Suharmoko, Aditya, "Govt Proposes to Reduced Fuel Subsidy", in: *The Jakarta Post* (16 August 2008), at: http://www.thejakartapost.com/news/2008/08/16/govt-proposes-reduced-fuel-subsidy.html (6 February 2012); Alfian; Suharmoko, Aditya, "Indonesia: Indonesia Eyes Reduced Fuel Subsidy", in: *The Jakarta Post* (17 August 2008), at: http://www.asianewsnet.net/print.php?id=892 (22 August 2008); "WRAPUP 1-Vietnam Hikes Fuel Price, Jakarta Bides Time on Subsidy Cut", in: *Alibaba.com* (24 February 2011), at: http://news.alibaba.com/article/detail/energy/100446652-1-wrapup-1-vietnam-hikes-fuel-price%252C.html (7 January 2012).

4.3.2 Malaysia

Malaysia, through subsidies to fuel retailers, has some of Asia's lowest petrol and diesel prices, with prices below market rates. In June 2008, prices of fuel and electricity were raised—petrol and diesel prices increased by 40 and 60 % respectively, while prices of electricity rose for households and commercial and industrial users by 18 and 26 %, respectively. The Malaysian government was expected to spend nearly US$14 billion (MR45 billion) on oil and gas subsidies (including tax relief and the gas subsidy for Petronas) in 2008 or about 7.5 % of GDP.[23] The government announced that beginning 1 September 2008, retail gasoline prices would be allowed to track global crude oil prices, with the subsidy fixed at 0.30 ringgit (US$0.09) per litre and pump prices adjusted monthly. By 2011, the fuel subsidy was estimated to increase to MR14 billion[24] due to high international crude oil prices thus causing an increase in the government's budget deficit.[25]

4.3.3 Thailand

As of 2011, Thailand has maintained its subsidy on diesel fuel; the State Oil Fund, which was established in 1979 to protect domestic fuel prices from a global fuel crisis, finances the subsidy. The fund imposes levies on certain key oil products when prices are declining and subsidises domestic fuel when global prices are rising. The fund also partly subsidises alternative fuels for vehicles, such as E-20, E-85, LPG and natural gas.

---

[23] "Malaysia Lifts 2008 Oil and Gas Subsidy Estimate to $14.24 Bn", in: *Alexander's Gas and Oil Connections* (8 May 2008), at: http://www.gasandoil.com/goc/news/nts82393.htm (21 August 2008); Bradsher, Keith, "Fuel Subsidies Overseas Take a Toll on U.S.", in: *The New York Times* (28 July 2008), at: http://www.nytimes.com/2008/07/28/business/worldbusiness/28subsidy.html (7 January 2012).

[24] "Malaysia's Fuel Subsidy to Rise", in: *The Straits Times* (10 March 2011), at: http://www.straitstimes.com/BreakingNews/SEAsia/Story/STIStory_643621.html (22 August 2011).

[25] For more information, see Bradsher, Keith, "Fuel Subsidies Overseas Take a Toll on U.S.", in: *The New York Times* (28 July 2008), at: http://www.nytimes.com/2008/07/28/business/worldbusiness/28subsidy.html (7 January 2012); Bradsher, Keith, "Prices of Food and Gas Take a Toll in Asia", in: *The New York Times* (23 July 2008), at: http://www.nytimes.com/2008/07/23/business/worldbusiness/23inflate.html (7 January 2012); Koswanage, Niluksi; Ahmad, Razak, "Najib and Malaysia's Fuel Subsidies", in: *Malaysia's Dilemma: Analysing Malaysia's Economic, Political and Social Dilemmas* (27 May 2011), at: http://malaysiasdilemma.wordpress.com/2011/05/27/najib-and-malaysias-fuel-subsidies/ (7 January 2012); "Malaysia's Fuel Subsidy to Rise", in: *The Straits Times* (10 March 2011), at: http://www.straitstimes.com/BreakingNews/SEAsia/Story/STIStory_643621.html (22 August 2011); "Najib: Fuel Subsidies to be Cut Gradually", in: *The Malaysian Insider* (17 May 2011), at: http://www.themalaysianinsider.com/mobile/malaysia/article/najib-fuel-subsidies-to-be-cut-gradually/ (4 October 2011).

LPG prices in the country are a problem area as prices continue to remain below international levels. Demand for LPG has also increased yearly by 10–15 % due to low fuel prices and increased vehicular conversion to LPG. Thailand, which was a net LPG exporter, became a net importer of the fuel in April 2008. In September 2011, the Thai energy minister announced that the government would borrow 10 billion baht from commercial banks to finance domestic fuel subsidies.[26]

4.3.4 Vietnam

Following several consecutive months of double-digit inflation—inflation was 26.8 % in June 2008—the Vietnamese government imposed price controls in March 2008 on ten 'essential commodities', including petrol, electricity, coal, water, cement, steel, school and hospital fees, and bus, rail and air travel. Unable to sustain subsidies on imported fuel—the government paid about US$500 million for oil subsidies during the first 5 months of 2008[27]—Vietnam raised domestic fuel prices by 36 % in July 2008.[28]

Although Vietnam is Southeast Asia's third-largest crude producer, it imports most of its petroleum product needs due to lack of refining capacity—its first refinery started commercial operations only in February 2009. The price controls imposed in March 2008 effectively postponed the deregulation of the Vietnamese oil industry. While gasoline prices have been nominally deregulated since mid-2007, importers still need government approval when setting end-user prices. State-owned companies (including Viet Nam Airlines Corp.), which account for 40 % of Vietnam's GDP, have complained about losses incurred due to these price controls. Companies in Vietnam were also facing difficulties in getting new loans to raise capital for petroleum imports. Some measures the government has taken to keep retail petrol prices stable are slashing import duties to zero and allowing petrol distributors to use the petrol price stabilisation fund to make up for losses.[29]

---

[26] For more information, see "B10 bn in Loans for Fuel Subsidy", in: *Diplomatsthailand* (29 September 2011), at: http://www.diplomatsthailand.com/top-stories/b10bn-in-loans-for-fuel-subsidy (8 October 2011); Master, Ammar, "Fuel Subsidies under Scrutiny in Thailand, Indonesia", in: *AutoBully* (10 August 2011), at: http://autobully.com/2011/08/fuel-subsidies-under-scrutiny-in-thailand-indonesia/ (7 January 2012); "Thai PTT Calls for End to Fuel Subsidies", in: *Reuters* (6 August 2008), at: http://www.reuters.com/article/2008/08/06/thailand-fuel-subsidies-idUSSP31001920080806 (7 January 2012).

[27] McCool, Grant, "Oil Price Forces Vietnam to Lift Controls", in: *The Brunei Times* (22 June 2008), at: http://www.bt.com.bn/en/node/44235/print (7 January 2012).

[28] Clark, Helen, "Vietnam: Coping with Skyrocketing Fuel Prices", in: *Inter Press Service* (1 August 2008), at: http://ipsnews.net/print.asp?idnews=43410 (7 January 2012).

[29] For more information, see Altman, Daniel, "Putting A Lasso on Inflation", in: *The New York Times* (29 July 2008), at: http://www.nytimes.com/2008/07/29/business/worldbusiness/29iht-glob30.1.14850266.html (7 January 2012); Clark, Helen, "Vietnam: Coping with Skyrocketing

4.3.5 Cambodia, Myanmar and Laos

The Cambodian government announced fuel subsidies of US$300 million for 2008,[30] or about 3.5 % of GDP, to control inflation due to rising oil prices. In Cambodia, fuel prices are generally unregulated and fuel subsidised indirectly (Jha et al. 2009, p. 9).

Laos can be considered to have small fuel subsidies. In 2008, the Lao government subsidised fuel when world oil prices reached historical highs. The government provided fuel subsidies of 300–400 kip (US$0.035–0.046) per litre of fuel.[31] With daily consumption in the range of ~1.2–1.3 million litres of fuel, the government paid about 520 million kip (US$60,100) a day or nearly US$20 million a year in fuel subsidies.[32] The fuel subsidy in Laos for 2008 was then less than 1 % of the country's GDP. When world oil prices dropped in early 2009 due to the global financial crisis, the Lao government maintained its high reference price for imported fuel in order not to lose the tax revenue from oil imports. With the continued rise of world oil prices in 2011, the government was considering fuel subsidies to curb inflation arising from the resultant increases in the prices of food and transportation.[33]

Myanmar unexpectedly reduced fuel subsidies in August 2007, resulting in fuel price increases that triggered violent public protests. Prices for CNG, promoted by the government for use in commercial vehicles, increased fivefold. The higher transportation costs, with bus fares and taxi charges doubling in urban centres, led to increases in the price of basic commodities such as food (rice, noodles, edible oil, meat, eggs, etc.). The urban poor were disproportionately affected, with manual workers and day labourers in cities earning less than US$2 a day having to pay more than half their wage

(Footnote 29 continued)
Fuel Prices", in: *Inter Press Service* (1 August 2008), at: http://ipsnews.net/print.asp?idnews=43410 (7 January 2012); McCool, Grant, "Oil Price Forces Vietnam to Lift Controls", in: *The Brunei Times* (22 June 2008), at: http://www.bt.com.bn/en/node/44235/print (7 January 2012); "Vietnam May Increase Fuel Prices to Cope With Rising Oil Costs", in: *Vietnam Business Finance* (16 July 2008), at http://www.vnbusinessnews.com/2008/07/Vietnam-may-increase-fuel-prices-to.html (21 August 2008); "Vietnam's First Refinery Debuts", in: *PVPro* (17 January 2011), at: http://www.pvpro.com.vn/en/news/detail/vietnam's-first-refinery-debuts-357.html (7 January 2012); "WRAPUP 1-Vietnam Hikes Fuel Price, Jakarta Bides Time on Subsidy Cut", in: *Alibaba.com* (24 February 2011), at: http://news.alibaba.com/article/detail/energy/100446652-1-wrapup-1-vietnam-hikes-fuel-price%252C.html (7 January 2012).

[30] "Cambodia's Fuel Subsidies to Cost 300 Mln USD in 2008", in: *Khmerization* (29 May 2008), at: http://khmerization.blogspot.com/2008/05/cambodias-fuel-subsidies-to-cost-300.html (7 March 2012).

[31] "Laos Raises Retail Fuel Prices", in: *China Radio International* (3 July 2008), at: http://english.cri.cn/3130/2008/07/03/1781s376508.htm (7 January 2012).

[32] Ibid.

[33] For more information, see "Lao Government Considers Fuel Subsidy", in: *Laovoices* (11 March 2011), at: http://laovoices.com/lao-government-considers-fuel-subsidy/ (7 January 2012); "Laos Raises Retail Fuel Prices", in: *China Radio International* (3 July 2008), at: http://english.cri.cn/3130/2008/07/03/1781s376508.htm (7 January 2012).

in travel costs. With more than 90 % of the population living in poverty, according to some economists based in Yangon, the drastic increase in fuel price was a sudden shock for the poor who had insufficient incomes to purchase even basic daily needs.

The Myanmar government's current policy is to provide fuel subsidies that cap retail petrol (gasoline) prices at K2,500 a gallon.[34] Customers can only buy a maximum of 4 gal of petrol at K2,500 a gallon during one fill-up. This policy has spawned a black market where the price of petrol is higher. In early 2011, the government rescinded the quantity restriction on diesel purchases. Diesel, which is used to run generators in times of power brownouts, can be bought in unlimited quantities. This rescinding of the quantity restriction on diesel purchases has been interpreted by some as official recognition of what was occurring in the black market for diesel.[35]

4.3.6 Brunei

Fuel subsidies in Brunei, which were 50 million Brunei dollars in 2004, quadrupled over the 2004–2007 period—the government reported that it spent US$151 million (202 million Brunei dollars) in fuel subsidies in 2007 or about 1 % of its GDP.[36] In 2008, Brunei maintained retail gasoline prices at US$0.40 (53 Brunei cents) a litre despite soaring international prices. Some residents suggest the reduction of fuel subsidies for residents to appreciate the scarcity value of oil, with a resident even commenting that it was odd that bottled mineral water was more expensive than oil in Brunei. Due to the fuel subsidies, Brunei has the highest energy use per capita and electricity consumption per capita among the ASEAN countries (Figs. 4.4 and 4.11).

As reflected in its GDP, Brunei as an oil exporter has benefited from the rise in world oil prices. A 2010 study by the national think-tank Centre for Strategic and Policy Studies (CSPS) reported that Brunei has 17 and 30 years supply of crude oil and natural gas, respectively, based on existing reserves. In 2011, the Minister of Energy stated that general fuel subsidies would continue and that the transition to

[34] Gaung, Juliet Shwe, "Fuel Meeting Ignores Black Market", in: *The Myanmar Times* (28 February-6 March 2011), at: http://www.mmtimes.com/2011/business/564/biz56403.html (7 January 2012).

[35] For more information, see Devine, Siobhan, "Analysis: Gas, Oil Get Myanmar Off Hook", in: *United Press International* (27 September 2007), at: http://www.upi.com/Business_News/Energy-Resources/2007/09/27/Analysis-Gas-oil-get-Myanmar-off-hook/UPI-81301190907679/ (7 January 2012); Gaung, Juliet Shwe, "Fuel Meeting Ignores Black Market", in: *The Myanmar Times* (28 February-6 March 2011), at: http://www.mmtimes.com/2011/business/564/biz56403.html (7 January 2012).

[36] "Brunei Tightens Belt as Fuel Price Subsidies Quadruple in 3 Years", in: *International Herald Tribune* (26 May 2008), at: http://www.iht.com/bin/printfriendly.php?id=13204982 (22 August 2008).

targeted or smart subsidies would only take place when there is adequate provision of public transport and further developments in the use of alternative energy.[37]

Table 4.3 summarises the indicative amounts of fossil fuel subsidies for select Southeast Asian countries for the 2004–2011 period.

4.3.7 Lessons

A reduction of fossil fuel subsidies is both an economic and political decision. Once subsidies are in place, it is politically difficult to remove or reduce them. Most governments are constrained by political considerations and concerns over potential unrest that may arise from such reduction. The experiences of ASEAN countries show the inefficient resource allocation effects of fossil fuel subsidies, the resulting fiscal burden on the government and their inefficiency in helping the poor since only a small proportion of the subsidy goes to the poor. Also, fuel subsidies often entail opportunity costs—money that could have been spent on education, health, infrastructure and other government programmes that entail longer-term benefits for the poor get directed to subsidies instead.

With subsidised oil prices, consumers do not face the true scarcity value of the resource and therefore have less incentive to conserve; similarly, firms have less incentive to develop and pursue alternative fuel and energy sources, too. Fossil fuel subsidies undermine the competitiveness of renewables and other energy technologies. At a subsidised fuel price, demand will be higher than what it would be if market forces were left to operate by themselves. This effect on demand is starkly apparent in the case of Brunei where energy subsidies have led to very high per capita energy consumption and are hastening the depletion of exhaustible resources. The demand effect can be quite significant in that some countries, such as Indonesia with oil and Thailand with LPG, shift from being net exporters to net importers of subsidised goods. With finite resources, greater consumption today means less availability of the resource for future generations. This intergenerational concern has been raised in Brunei and Malaysia (depleting Petronas' profit for fuel subsidy), which are net exporters of oil, a non-renewable resource.

In addition to imposing fiscal burden on the government, fuel subsidies, as experienced by Vietnam's state-owned corporations, can also make it more difficult for oil and electricity companies to pursue their commercial objectives independent of the government's social policies. Political influence in the setting of prices and the resulting regulatory uncertainty can hinder private investment in these sectors. In the Philippines, a major impetus, aside from the perceived high

[37] For more information, see "Brunei Tightens Belt as Fuel Price Subsidies Quadruple in 3 Years", in: *International Herald Tribune* (26 May 2008), at: http://www.iht.com/bin/printfriendly.php?id=13204982 (22 August 2008); Masli, Ubaidillah, "Fuel Subsidy is about Well-being", in: *The Brunei Times* (2 February 2011), at: http://www.bt.com.bn/news-national/2011/02/02/fuel-subsidy-about-well-being (7 January 2012).

Table 4.3 Amount of fossil fuel subsidies in select Southeast Asian countries, 2004–2011

Country	Year	Subsidy	
		Type	Amount
Brunei	2004[a]	Fuel	50 million Brunei dollars
	2007[b]	Fuel	US$151 million or 202 million Brunei dollars (~1 % of GDP)
Cambodia	2008[c]	Fuel	US$300 million (3.5 % of GDP)
Indonesia	2008[d]	Fuel	US$20 billion or 180.31 trillion rupiahs (4 % of GDP)
	2009[e]	Electricity	US$6.5 billion or 60.4 trillion rupiahs
	2010[f]	Fuel	Expected to be 8 % of public expenditure (government budget)
	2011[g]	Fuel	Expected to be 8 % of public expenditure (government budget)
Laos	2008[h]	Fuel	US$20 million (<1 % of GDP)
Malaysia	2008[i]	Oil and gas	Estimated at US$14 billion or MR45 billion (7.5 % of GDP)
	2011[j]	Fuel	Expected to be MR14 billion
Thailand	2011[k]	Fuel	nd[m]
Vietnam	2008[l]	Oil	US$500 million[n]

Source Various (Provided separately for each country in the table)

[a] Brunei Tightens Belt as Fuel Price Subsidies Quadruple in 3 Years", in: *International Herald Tribune* (26 May 2008), at: http://www.iht.com/bin/printfriendly.php?id=13204982 (22 August 2008)

[b] Ibid

[c] Cambodia's Fuel Subsidies to Cost 300 Mln USD in 2008", in: *Khmerization* (29 May 2008), at: http://khmerization.blogspot.com/2008/05/cambodias-fuel-subsidies-to-cost-300.html (7 March 2012)

[d] Alfian; Suharmoko, Aditya, "Govt Proposes to Reduced Fuel Subsidy", in: *The Jakarta Post* (16 August 2008), at: http://www.thejakartapost.com/news/2008/08/16/govt-proposes-reduced-fuel-subsidy.html (6 February 2012); Bradsher, Keith, "Fuel Subsidies Overseas Take a Toll on U.S.", in: *The New York Times* (28 July 2008), at: http://www.nytimes.com/2008/07/28/business/worldbusiness/28subsidy.html (7 January 2012)

[e] Alfian; Suharmoko, Aditya, "Govt Proposes to Reduced Fuel Subsidy", in: *The Jakarta Post* (16 August 2008), at: http://www.thejakartapost.com/news/2008/08/16/govt-proposes-reduced-fuel-subsidy.html (6 February 2012)

[f] Fiscal Policy Office, Ministry of Finance (Indonesia) 2011: 14, at: http://www.globalsubsidies.org/files/assets/ffs_gsiunepconf_sess2_askolani.pdf (7 January 2012)

[g] Ibid

[h] "Laos Raises Retail Fuel Prices", in: *China Radio International* (3 July 2008), at: http://english.cri.cn/3130/2008/07/03/1781s376508.htm (7 January 2012)

[i] "Malaysia Lifts 2008 Oil and Gas Subsidy Estimate to $14.24 Bn", in: *Alexander's Gas and Oil Connections* (8 May 2008), at: http://www.gasandoil.com/goc/news/nts82393.htm (21 August 2008); Bradsher, Keith, "Fuel Subsidies Overseas Take a Toll on U.S.", in: *The New York Times* (28 July 2008), at: http://www.nytimes.com/2008/07/28/business/worldbusiness/28subsidy.html (7 January 2012)

[j] "Malaysia's Fuel Subsidy to Rise", in: *The Straits Times* (10 March 2011), at: http://www.straitstimes.com/BreakingNews/SEAsia/Story/STIStory_643621.html (22 August 2011)

[k] "B10bn in Loans for Fuel Subsidy", in: *Diplomatsthailand* (29 September 2011), at: http://www.diplomatsthailand.com/top-stories/b10bn-in-loans-for-fuel-subsidy (8 October 2011)

[l] McCool, Grant, "Oil Price Forces Vietnam to Lift Controls", in: *The Brunei Times* (22 June 2008), at: http://www.bt.com.bn/en/node/44235/print (7 January 2012)

Abbreviations GDP = gross domestic product; *nd* = not determined

[m] The government announced in September 2011 that it intends to borrow 10 billion baht from commercial banks to finance domestic fuel subsidies

[n] For the first five months of 2008

electricity prices, for the privatisation of the generation and transmission assets of the government-owned NPC as well as the overall restructuring of the power industry was the substantial losses then being incurred by NPC. In addition to the usual efficiency issues in relation to government operation, these losses were partly due to the political influence in rate setting. The government, wanting to appease the populace, mandated reductions or caps in power rate increases that caused substantial losses to the firm. Since these NPC loans were contingent liabilities for the government, the losses also contributed to the fiscal deficit.

If the objective of fuel subsidies is to help the poor cope with high fuel prices, then broad fuel subsidies are an inefficient and costly mechanism—a substantial portion of the subsidy goes to richer consumers, who are able to afford cars, air conditioners, televisions, refrigerators, etc., and energy-intensive industries that tend to be capital-intensive. In Indonesia, for instance, official estimates indicate that about two-thirds of its fuel subsidy accrues to the top (richest) 40 % of the income distribution, with only 15 % going to the bottom (poorest) 40 % of the distribution.[38]

Broad fuel subsidies inadvertently act as an input subsidy for production sectors that directly or indirectly use the subsidised fuel. The price distortion arising from such fuel subsidies biases resource allocation away from labour-intensive industries. In lower-income countries, employment generation generally helps the middle- and lower-income classes. In general, there is no clamour from the production sectors of Southeast Asian countries for an explicit fuel or electricity subsidy through subsidised rates specific to production firms, though concerns regarding reliable energy supply at 'reasonable' prices remain. Firms generally expect their governments to ensure the timely installation of energy infrastructure (generation plants, transmission assets, regulatory framework, etc.) and provide a smooth-functioning economic environment that reduces the cost of doing business. In most Southeast Asian developing countries, such initiatives would include reducing corruption; provision of roads, ports, airports and other transport infrastructure to reduce transport costs; and strengthening institutions including the enforcement of contracts, simplifying government transactions and procedures, and reducing the time needed to process government permits and licenses.

More targeted income support for poor families would be preferable to broad fuel subsidies. The U-Save Rebate scheme in Singapore, for instance, is a well-designed income support programme for lower- and middle-income households. The scheme, however, may not be replicable in lower-income countries where most of the poor are in informal housing markets and rural areas, and the nature and characteristics of poverty are different. Examples of existing alternative social assistance programmes include the conditional cash transfer (CCT) programmes in the Philippines and Indonesia.[39] In the Philippines, subsequent to the surge in

---

[38] Alfian; Suharmoko, Aditya, "Indonesia: Indonesia Eyes Reduced Fuel Subsidy", in: *The Jakarta Post* (17 August 2008), at: http://www.asianewsnet.net/print.php?id=892 (22 August 2008).

[39] Though CCTs can reduce poverty in the short term through cash transfers and in the long term through conditionalities on human capital development (usually health and education),

world oil prices in 2008 and a revenue windfall in the form of value-added tax (VAT), the government has implemented several short-term income support schemes for poor families, including one-time P500 grants to households consuming less than 100 kWh a month (effectively a cash transfer, with electricity consumption as an identifier for being poor), cash aid to the elderly and CCT programmes (cash grants conditional on health centre visits and school attendance by children) of the Department of Social Welfare and Development (DSWD).

With the rise in fuel prices, some governments have been confronted with political demands for 'mitigating measures' from negatively affected sectors, such as the transport sector. Such demands usually come from lower-income groups, such as jeepney drivers in the Philippines who provide public transportation mostly to lower- to middle-income commuters. In response, for a few months in 2011, the Philippine government provided a monthly fixed-amount subsidy to some 140,000 jeepney drivers in the form of fuel assistance smart cards—with a load of P1,050 (~US$24)—that could be used when filling up at petrol stations; cost to the government was in the range of several hundred million pesos: P450 million (~US$10 million). Arguably, were the administrative cost of the short-lived programme also considered, it might have been better in the long term had the government used the money instead for road repair and improving traffic flow, which in the long run would have benefited a greater number of commuters through transport cost savings.[40] Governments should therefore undertake some cost-benefit analyses, in view of the social objectives involved, of such 'token' subsidy programmes vis-à-vis other uses of money.

4.4 Socioeconomic Impacts: Macroeconomic Factors

4.4.1 *Effects on the Economy*

With rising oil prices, oil-exporting countries, such as Brunei, are expected to gain while oil importers are expected to confront rising economic and social costs, at least in the short run before the economy can adjust to the higher oil prices. The macroeconomic effects of rising oil prices on net importers of crude oil are mounting inflation, slower growth, higher trade deficits and weaker fiscal positions. Weaker fiscal positions may result if the slower growth in economic output yields lower tax revenues as the government increases social spending and

(Footnote 39 continued)
they are not necessarily good instruments for dealing with crisis situations and transient poverty. CCTs are better suited for dealing with structural poverty (Handayani and Burkley 2010, pp. xxvii–xxx).

[40] Transport cost savings can include savings in transport time, fuel, vehicle maintenance and environmental damages from less air pollution.

undertakes expenditures to stimulate the economy. Oil-importing countries in the region like Singapore can, however, gain from trading with oil exporters that benefit from higher oil prices, thus tempering the negative impacts of the oil price increases. In high-growth countries, such as China and India, the negative (partial) macroeconomic effects of rising oil prices may be outweighed by the positive effects of high growth rates due to rising external and internal demand, leading possibly to net positive effects on growth and employment, trade balance and fiscal positions while the impact on inflation is inconclusive.

Together with the general increase in global food prices, the increase in oil prices has led to an increase in inflation, compelling monetary authorities to move towards more restrictive monetary policies to temper rising inflationary expectations. The oil price increases have led to demands for wage increases by workers to cover higher living costs, fare increases by transport operators and higher prices of goods by firms having to cover increased production costs. The low interest rates in the US, however, restrict Asian central banks from raising their interest rates as it may lead to rapid appreciation of their currencies. Stronger currencies, while making oil imports cheaper in terms of the local currency, could make their exports less competitive. Furthermore, demand in the international market is expected to weaken in the face of the global economic slowdown resulting from the global credit market crisis due to the US subprime mortgage problem, the European debt crisis, and higher energy and food prices. Higher interest rates can also be expected to lead to slower economic growth.[41] Also, higher inflation in Asia could be transmitted outside the region through exports to countries like the US.

While there are other structural factors underlying the global surge in food prices, econometric studies have shown that a rise in oil prices has a statistically significant positive impact on food prices (ADB 2008). The rising price of oil, closely associated with the price of natural gas, is reflected in increases in fertiliser prices. There are also higher costs for energy (fuel and electricity) for pump irrigation systems, farm equipment (such as tractors, harvesters and threshers) and fishing boats. Transportation costs, and hence distribution costs, for agricultural inputs and output also go up—all of which contribute to a general hike in the prices of agricultural produce.

Price inflation has not been confined to food and energy, but has spread to the core inflation rate that excludes food and energy. The higher import value of oil and weaker exports negatively affect the balance of trade for a country—a trade deficit is usually interpreted as a transfer of wealth to other countries. The country's trade balance could improve if domestic biofuels production and other

[41] By the third quarter of 2011, due to the European debt crisis that was triggered by the debt problems of Greece and its spillover effects on the growth of the global economy, inflation was no longer a major concern for most monetary authorities. Given the expectation that global demand will decline, increases in interest rates to temper inflationary expectations are generally viewed as not warranted. Monetary authorities, in this case, are more likely to hold interest rate increases in order to stimulate domestic demand to compensate for a drop in exports.

indigenous sources are able to replace some of its oil imports. Energy exporting countries, such as Indonesia with its coal and palm oil, are partially insulated from this phenomenon.

Oil importers may indirectly gain from rising oil prices through trade in other goods and services with oil exporters, such as the Middle East countries, Russia and Venezuela, as exports to oil-rich countries are likely to increase. Malaysia, for example, can expect more tourists from the Middle East. Countries like Indonesia and the Philippines, which are major labour exporters to the Middle East, can also take advantage of the expected economic boom in the region with an increased demand for foreign labour in these countries. The windfall profits of the oil-rich countries may also be reinvested in the form of foreign direct investment in other countries, including those in the Southeast Asian region, or recirculate as assistance to Muslim-dominated areas, such as Southern Philippines and Indonesia.

The cost of shipping, from Asia to the US or Europe for instance, has gone up too. The effects of increases in shipping costs are said to be similar to those of a tariff on trade. A study by the Canadian investment bank, CIBC World Markets, in May 2008 estimated that the increase in shipping costs, prior to the 2008 global financial crisis, was on average equivalent to a 9 % tariff on trade.[42] Higher oil prices and resulting increases in transportation costs also affect international trade configurations through the so-called 'neighbourhood effect', wherein production is generally moved closer to component suppliers and to consumers to reduce transportation costs. The trend towards greater global trade in goods and services is expected to persist though, as transportation cost is just one factor in the decision of companies on where to locate a new factory or from whom to buy a product. Exchange rates, labour costs, availability of skilled workers, government regulations and political stability are some of the other factors that determine trade and investment flows. Higher transportation costs are more likely to affect industries that produce goods that are expensive to ship relative to their product price and may induce greater regional trade wherein distances are shorter.

4.4.2 Taxation

Tax revenue from oil has become a major source of government revenue for some ASEAN countries especially with recent surges in oil prices. There has been political clamour in these countries for a reduction or suspension of taxes on oil. In the Philippines, the VAT has been a source of revenue windfall in the form of increased tax collection from oil products. There has been a proposal to shift taxation from an ad valorem VAT system on petroleum products to a specific tax, thus fixing tax revenue per unit of oil irrespective of price changes. Another

[42] Rohter, Larry: "Shipping Costs Start to Crimp Globalization", in: *The New York Times* (3 August 2008), at: http://www.nytimes.com/2008/08/03/business/worldbusiness/03global.html (7 January 2012).

proposal is to reduce VAT rates on oil products and increase the rates for other products, thus distorting the universality of the VAT system. The Philippine government has, in response, adopted the position that the current VAT system, at 12 % rate on all products but zero-rating for some agricultural products, is necessary to finance short-term income support for the poor and its various long-term programmes, such as infrastructure, health and education investments.

4.4.3 Energy: Regional Trade and Cooperation

Regional trade in energy has fostered greater country interdependence in Southeast Asia. Examples of trade flows and cross-border investments in energy within Southeast Asian countries are:

- Indonesia exports natural gas to Singapore. The ASEAN is also considering the feasibility of a 1,000 km natural gas pipeline from Indonesia's Natuna gas field to Malaysia and Thailand that could further be routed to Brunei, Sarawak and the Philippines.
- Myanmar is one of Thailand's biggest sources of energy, with Thailand's biggest private power producer sourcing all of its natural gas from Myanmar's fields. The Yadana-Ratchaburi gas pipeline links the two countries.
- Thailand purchases hydropower from Laos.
- The Philippines buys coal from Indonesia.
- Golden Hope Plantations, a Malaysian state-run palm oil company, has invested in biofuels and has oil palm plantations in Indonesia.
- PTT Exploration and Production, a state-controlled Thai company, has contracts to explore and develop oil and gas fields in Vietnam, Myanmar and Indonesia.
- Thailand and Myanmar plan to jointly build hydropower plants and dams.
- Although it does not have its own oil and gas resources, Singapore, through the Singapore Petroleum Company, is active in upstream exploration and production of petroleum in Indonesia, Vietnam and Cambodia. In addition, it also provides support services, such as the construction and repair of drilling rigs and service vessels.

With relatively undeveloped hydropower potential, the Greater Mekong Subregion (GMS; including Cambodia, China, Laos, Myanmar, Thailand and Vietnam) could provide cost-effective energy access through cross-border hydroelectricity sales. The Asian Development Bank (ADB) is supporting a project that aims to connect power generation plants through a regional network of cross-border interconnections and wholly competitive buying and selling markets.[43]

---

[43] Sovacool, Benjamin K.; D'Agostino, Anthony, "Greater Mekong Subregion: Turning Water into Gold", in: *Asian Trends Monitoring Bulletin* (republished 27 June 2011), at: http://siew.sg/node/557 (7 January 2012).

Although power trading does entail energy security risks when compared to trading in energy sources (e.g. coal, oil and natural gas), there are benefits to the former, too. It may be cheaper for some countries to import electricity from neighbouring countries with excess capacity than to build new domestic capacity. For the exporting country, the local customer base may be too small and the export market for the electricity may make its investments in generating capacity financially viable.

4.4.4 The Global Financial Crisis and Recovery

Generally, the 2008 global financial crisis led to a slowdown in the economic growth of ASEAN countries in 2008 and 2009 (Fig. 4.12). Most impacted were the more open and export-dependent economies of Singapore, Brunei, Malaysia and Thailand that had negative real GDP growth rates or economic contraction in 2009. For these countries, the negative effects of the drop in non-oil export demand were to some extent cushioned by the fall in energy prices. For oil export-dependent Brunei, on the other hand, economic performance was positively linked to oil prices. However, by 2010, the ASEAN economies, showing resilience to the global exogenous shock, had recovered with real GDP growth rates near pre-crisis levels. The buoyancy of the recovery was more marked for Singapore, which posted a 14.5 % real GDP growth in 2010 following a 0.8 % contraction in 2009.

4.5 Socioeconomic Impacts: Microeconomic Factors

Economic agents, both energy producers and consumers, respond to changes in energy prices and adapt in the longer run. In general, if the price of a product goes up, demand goes down and consumers shift to substitute goods. In the longer run, the economy adapts to a perceived permanent change in relative prices, with a greater scope for adjustment on both the supply and demand sides; the price change also influences the direction of associated technological developments.

4.5.1 Supply Response

Although higher oil prices have made renewable, alternative green energy options attractive, most forms of renewable energy at present—from solar panels to wind farms—rely on subsidies for financial viability. Still, the huge scale of the energy market provides opportunities for alternatives to prove themselves at the margin

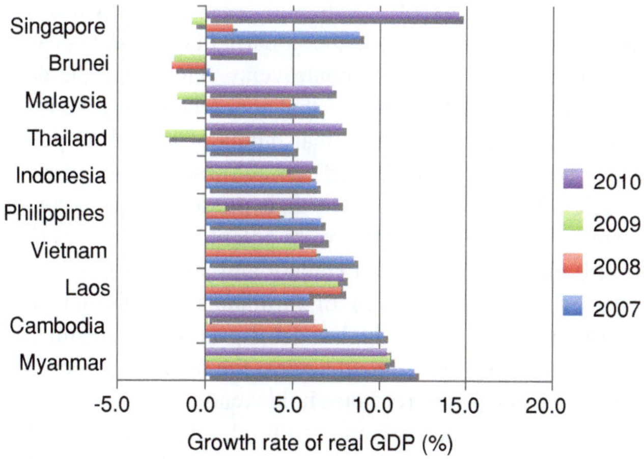

Fig. 4.12 Growth rates of real GDP, per cent, 2007–2010. *Source* ADB 2011, p. 173. *GDP* gross domestic product

and move to mainstream commercial scale as their costs, through technological improvements, decline.

4.5.1.1 Biofuels

Some ASEAN countries have set biofuel targets for their transport sector through renewable energy fuel standards. The increase in global food prices due to greater demand for biofuels and the recognition that it takes considerable energy to produce ethanol (e.g. corn for ethanol in the US)[44] has led to a search for more efficient non-food sources of biofuels, including agricultural waste. Environmental concerns related to raising crops for biofuels are: fuel competing with food for cropland (e.g. corn is better used as livestock feed than fuel) and the usual agricultural concerns, such as deforestation and haze resulting from land clearing, fertiliser use and agricultural runoff, water use, and crop variability due to weather and its resulting effects on the volatility of energy prices.

Biofuel production can be beneficial if marginal lands are used for the biofuel crops, competition with food crops is reduced and technologies converting crops to biofuel are relatively efficient in energy input–output terms. In climate change mitigation terms, however, biofuels can have adverse effects if the expansion of

[44] The ratio of how much energy is used to make ethanol to how much energy it delivers is known as the energy balance. An estimate given by the National Renewable Energy Laboratory (NREL) suggests that a British thermal unit (Btu) of fossil energy is consumed in producing and delivering corn ethanol that results in 1.3 Btu of usable energy in the fuel tank; see Meigs, James B., "The Ethanol Fallacy: Op-Ed", in: *Popular Mechanics* (18 December 2009), at: http://www.popularmechanics.com/cars/alternative-fuel/biofuels/4237539 (7 March 2012).

farmlands entails the clearing of forests and subsequent release of greenhouse gases (GHGs). Also, the matter of whether a particular biofuel is carbon-negative, -neutral or -positive is still subject to controversy. Therefore, the development of low-input high-diversity (LIHD) biofuels that are carbon-negative is being actively researched. LIHD biofuels require minimal inputs of irrigation, pesticides or fertilisers and can be grown on degraded agricultural land in plots containing a mixture of species. LIHD biofuels, when their production on a commercial scale becomes economically viable, could serve as potential alternatives to conventional first-generation biofuels, such as corn ethanol and soybean biodiesel.

When considering the conversion of farmlands for biofuel cultivation, the increase in farm incomes due to higher crop prices as a result of the greater demand for biofuels should be weighed against the negative effects of higher food prices on farm households—poorer households tend to spend a greater proportion, nearly one-third to half, of their incomes on food. Also, government subsidies for biofuel production are likely to exacerbate associated efficiency losses and its negative social and environmental effects. A difficulty in the selection of crops for biofuels is illustrated by the case of jatropha.

Jatropha, an inedible but hardy plant that grows on marginal land, was considered in the mid-2000s to have the potential to revolutionise biofuels.[45] In Southeast Asia, there was considerable interest in planting jatropha for biofuel in Indonesia, Myanmar[46] and the Philippines,[47] and currently in Thailand. It turned out though that marginal lands for jatropha also led to marginal yields—the plant requires a lot of water, especially when grown in arid and semi-arid areas, thus putting stress on water catchment basins. While current scientific evidence does suggest that jatropha grows more productively on higher quality land with more rainfall or irrigation, planting the biofuel in such areas will lead again to tradeoffs between food crops and those for fuel. Faced with such a tradeoff and uncertain income from biofuels, farmer households would put priority on food crops. Questions also persist on the economic and financial viability of large-scale jatropha plantations that will be required in order to have a significant effect on world fuel markets. For jatropha to be grown on an industrial scale, the plant will have to be domesticated (similar to earlier developments in the cultivation of

---

45 "Biodiesel Wonder Plant Could Spell Doom for Kenya", in: *The East African* (26 July–1 August 2010), at: http://naturekenya.org/sites/default/files/EastAfrican%2026July-01Aug2010.pdf (20 August 2011); "BP Gives Up on Jatropha for Biofuel", in: *The Wall Street Journal* (17 July 2009), at: http://blogs.wsj.com/environmentalcapital/2009/07/17/bp-gives-up-on-jatropha-for-biofuel/tab/print/ (7 January 2012); Luoma, Jon R., "Hailed as a Miracle Biofuel, Jatropha Falls Short of Hype", in: *Guardian Environment Network* (5 May 2009), at: http://www.guardian.co.uk/environment/2009/may/05/jatropha-biofuels-food-crops (7 January 2012).

46 Reports suggest that in 2005 the Myanmar government ordered the planting of jatropha but did not build adequate infrastructure to process them into biofuel. The seeds were then left to rot.

47 A large-scale jatropha plant project launched by the Arroyo administration lost PhP1 billion (around US$23 million) as cost incurred for planting jatropha on nearly 4,000 ha of land. The plant oil turned out to be not commercially viable; see "Jatropha Project Bombs", in: *Philippine Daily Inquirer* (7 November 2011): A9.

rubber trees) and its oil yields enhanced through plant breeding or genetic manipulation. Since commercial viability calls for the conversion of huge areas of land for jatropha cultivation, there are also concerns about its effects on biodiversity. In some parts of Australia, jatropha has in fact been declared an invasive plant.[48]

Biofuels of interest in Southeast Asia are mainly first-generation biofuels from food crops. Examples include coconut oil for coco diesel and sugarcane for bioethanol in the Philippines, and palm oil for biodiesel in Indonesia, Malaysia and Thailand. Singapore, without much agricultural land, aims to participate in the green development in Asia by focusing on the development of second-generation biofuels (from biomass or non-edible feedstock), solar energy and other alternative energy technologies, and encouraging research and development (R&D) and pilot plants in Singapore. An economic rationale for subsidies for R&D in alternative energy technologies is that innovators in energy markets cannot capture the full benefits of their innovations and would therefore tend to underinvest without the subsidies. Singapore has compensated for its lack of natural resources by the creation and use of human capital for sustainable development.

4.5.1.2 Renewable Energy

The conventional fuels for power generation are oil, natural gas and coal. With the increased oil prices, the prices of natural gas and coal have also increased but to a lesser extent. Natural gas is also seen as less polluting than coal. The high price of oil and opposition from local environmental groups to the use of coal has also resulted in growing use of natural gas for power generation in the Southeast Asian region.

In Southeast Asia, renewable energy—particularly solar, wind and hydropower—is usually used for rural off-grid electrification with substantial government subsidy. Scaling up of renewable energy use for base load power generation though has not yet taken hold widely. Examples of grid supply from renewable sources include the 8 MW Bangui Wind Power Project in coastal northern Luzon in the Philippines and a 950 kW 2-ha solar power operation in Mindanao in southern Philippines; both projects sell their output to local distribution utilities.

Hydroelectric dams, such as those in the GMS, have also been constructed in Southeast Asia to meet increased energy demand. However, in the case of such projects, the social cost of population resettlement and environmental effects, such as the effects on fish migration and biodiversity, have to be weighed against the social benefits of increased power availability and lower power rates. An important equity issue is the distribution of benefits, usually accruing to higher-income urban

[48] "Climate Change: Jatropha—Not Really Green", in: *IRIN: Humanitarian News and Analysis* (23 March 2011), at: http://www.irinnews.org/printreport.aspx?reportid=92267 (7 January 2012).

areas with higher demand for electricity, and costs, the brunt of which is borne by rural residents.

The use of modern renewable energy in Southeast Asia is, however, limited by the state of renewable energy technologies and their cost competitiveness, as illustrated by the controversies surrounding feed-in-tariffs (FIT)—guaranteed prices at which renewable energy suppliers will be paid for the electricity that they produce—for grid supply in the Philippines. Electricity generated by FIT-certified renewable energy sources is classified as 'must dispatch', for which a universal levy—a uniform per kWh charge or FIT-allowance—is paid by electricity consumers. Industry groups have been vocal about concerns that FITs and the associated universal levy will result in higher prices of electricity, which will then negatively affect their competitiveness.

The Philippine Department of Energy has proposed an installation target of 760 MW to be covered by the FIT rates over three years.[49] The proposed FIT rates, guaranteed over a 20-year period, and capacity allocation targets for specific renewable energy technologies are given in Table 4.4. In 2014, when all renewable energy projects are expected to have gone online, the FIT-allowance or universal charge is estimated to be 12.75 centavos/kWh (PhP 0.1275/kWh), or about 2–3 % of current generation charges. With current generation charges being around 5–6 pesos/kWh,[50] proposed FIT rates for the more expensive wind, ocean and solar technologies are nearly 2–3 times that of existing generation charges, which cover electricity generated not only from fossil fuels but also renewable resources, such as hydro and geothermal power. Only power generated from new investments will be covered by these FIT rates.

The controversy arising from the opposition to FIT and the associated increase in electricity prices by some business and industry groups has delayed its implementation in the Philippines. The resulting uncertainty has hampered investment in renewable energy projects by local and foreign investors alike. Indicating significant investor interest and awaiting the implementation of the FIT are 236 signed service contracts with a projected capacity of 2,822 MW, and about 400 pending applications for another 6,000 MW capacity.[51] A proposal is the phased implementation of renewable energy technologies, with the cheaper biomass and hydro options being commissioned first and the more expensive technologies, such as wind and solar energy, being delayed until advances in their technologies lower associated costs. Another proposal is the auction of the limited installation allotments of 760 MW for renewable energy sources as an alternative to the imposition

49 This additional capacity is about 5 % of the existing total installed capacity in the Philippines, which was estimated at around 15,610 MW in 2004; see DOE Portal, the Philippine Department of Energy, at: http://www.doe.gov.ph/statistics/power.htm (7 January 2012).

50 This figure applies to the Meralco franchise area, which is the largest distribution utility in the country covering the National Capital Region (NCR) and surrounding areas.

51 Remo, Amy R., "Implement 'Green' Power Projects in Phases", in: *Philippine Daily Inquirer* (29 September 2011): B2.

Table 4.4 Proposed FIT rates and capacity allocation, The Philippines

Renewable energy resource	Proposed FIT (PhP/kWh)	Proposed capacity allocation[a] (MW)
Biomass	7.00	250
Ocean	17.65	10
Run-of-river hydro	6.15	250
Solar (ground mounted)	17.95	50
Wind	10.37	200

Sources: Energy Regulatory Commission 2011, (Energy Regulatory Commission (the Philippines), *Notice of Public Hearing: Notice of Proposed Rule-Making*, Pasig City, Philippines (4 August 2011), at: http://www.erc.gov.ph/pdf/FIT_ERC%20CASE%20NO%20%202011-006%20RM.pdf (7 January 2012).); Remo 2011 (Remo, Amy R., "Gov't Urged to Defer Renewable Energy Projects", in: *Philippine Daily Inquirer* (12 August 2011), at: http://business.inquirer.net/12235/gov't-urged-to-defer-renewable-energy-projects (23 February 2012))

FIT feed-in-tariffs, *PhP* Philippine Peso, *kWh* kilowatts hour, *MW* megawatts

[a] Based on the Department of Energy's submission of 'revised and certified' installation targets. The initial target of 830 MW was reduced based on a technical study, which showed that the grids and transmission facilities are only capable of handling an additional 760 MW

of FIT. Under the second proposal, the proposed FIT rates would serve as price ceilings and lower tariffs could result from an auction system.

4.5.2 Demand Response

Some sectors that are vulnerable to high oil prices are agriculture (fertiliser prices go up), fishing (fuel prices increase) and construction (the metal and cement industries are energy-intensive). There is greater incentive to pursue energy efficiency through the use of more efficient lighting, development of more efficient appliances and energy labelling, and more investments in better-designed mass transit systems.

Travel patterns change with fuel price changes. It has been observed that consumers adjust to high gasoline prices by driving private vehicles less, driving shorter distances, and making greater use of public transportation (buses and water ferry systems) and mass transit systems, contributing to less traffic congestion in urban areas. There is also a shift in demand towards more fuel-efficient engines and smaller cars. In the Philippines, for instance, demand for motorcycles has greatly increased in the past years (coupled with an increase in traffic accidents involving motorcycles). In Vietnam, where motorcycles are a common mode of transport, there is a return to the use of bicycles.

High transport costs may lead in the long run to higher densities in urban areas where most of the jobs are. Transportation costs to and from work for the average worker in developing countries can be a significant portion of their wages. Some workers, like those in factories and construction, may then prefer live-in arrangements. Even for middle-income families, it is expected that the choice of housing location will be affected by distance from work.

4.6 The Future: Economic Policies and Options

4.6.1 Energy Poverty, Greener Energy and Rural Development

Energy poverty for the rural poor manifests at varied levels. In the face of high energy prices, some of the poor are shifting from kerosene and LPG to charcoal and fuelwood, which exposes them to serious health risks from indoor pollution due to the use of biomass fuel. In developing countries, a significant portion of the lower-income groups does not have access to electricity, as in rural areas with dispersed households and low demand for power, it usually is not economical to provide grid-supplied electricity. Renewable energy in these countries is frequently used for rural missionary off-grid electrification projects, often involving solar power and mini-hydro systems. These projects usually require government subsidies or foreign assistance through aid programmes.

The emergence of biofuels and the accompanying increase in demand for crops may provide an opportunity for rural development and poverty alleviation through agricultural expansion. At the level of local communities where crops are raised, issues such as global warming, GHGs and carbon emissions may be distant considerations and primary concern likely to focus on issues such as local employment and household incomes. Biofuel production may, therefore, have a positive impact on rural communities if it involves small-scale farmers and requires lower-skilled poor workers in rural areas.

Though targeted aid and income support for poor families to cope with rising oil prices are preferable to broad fuel subsidies, such measures are perceived only as short-run solutions. Public consultations reveal that what matters most to the poor are employment and job creation that generates incomes so that they too can participate in the market economy to exchange goods and services. Consumers would not be as concerned about high prices of electricity and gas if they have the incomes to purchase these goods. Although the poor in rural areas may subsist on their own supply of food, they may have insufficient funds for clothing, health services, books for school-aged children and other goods that have to be bought from the market. More permanent benefits to the poor can, therefore, only be attained through long-term investments on human capital, such as education and health. Overall economic growth will be necessary too to ensure a larger economic pie for redistribution.

4.6.2 Energy Efficiency, Conservation and Diversification

As the increase in oil prices is expected to be a more permanent shift and with the climate change issue at the forefront, greater demand for alternative greener energy is likely to continue. With the reliance of the world economy on fossil fuels, oil though is not likely to be replaced in the near future. Oil exploration still continues and alternative oil sources, such as oil sands are being developed.

However, higher oil prices will provide strong incentives for energy conservation and efficiency and the development of alternatives to fossil fuel. Economic growth leads to higher demand for energy, usually met by conventional fossil fuels; however, if the increased energy demand leads to higher fossil fuel prices, then the relative price of alternative energy is reduced which can provide an incentive for the development and use of renewable energy sources. Fossil fuel subsidies and other government policies that distort price signals will simply impede the long-run adjustment of the economy to the scarcity value of energy.

Energy conservation and efficiency can have significant payoffs in the short and long terms[52] since the resulting reduction in energy demand can shed the need for new capital investment in power generation. The rate of adoption of alternative energy will depend on the price of alternative energy, including renewable energy, relative to that of conventional fossil fuels, government energy policies, and technological developments. A current problem with renewable and other alternative energy is their commercial viability, especially in scaling up these technologies for commercial mainstream operations.

Renewable energy can reduce dependency on imports and diversify a country's energy supply, thus reducing the vulnerability of the country to energy supply disruptions. The production of renewable energy can also reduce the levels of local pollution. For off-grid remote rural areas, renewable energy, such as solar and mini-hydropower projects, are sometimes the least-cost option. The promotion of renewable energy has been in the pursuit of 'greener' economic growth.

4.6.3 Promotion of Alternative Technologies: Subsidies

The policy objectives of subsidising renewable energy technologies include enhancing energy security by increased diversification of energy supply, the environmental goal of mitigating climate change, and the social goal of missionary electrification. Subsidies can take the form of price supports (e.g. FITs), tax incentives (e.g. income tax

[52] A major barrier to energy conservation and efficiency improvements arises when these projects entail large capital investments but have long payback periods. Even at a smaller scale, poorer households in countries with imperfect capital markets may find energy-saving bulbs too steep in price compared to that of incandescent bulbs even though the net savings over the life of the energy-saving bulb is higher.

holidays, exemption from VAT, duty-free importation of equipment and machinery, investment tax credits, accelerated depreciation, etc.) and loan guarantees.[53]

An argument for subsidies for alternative technologies, such as solar panels and wind turbines, is that the more these alternative technologies are installed or used, the technologies would move down the learning curve and eventually become cheaper and achieve economies of scale. Hence, subsidies, for example for solar panel and wind turbine installation, are needed to stimulate demand. At the stage where the cost of production is sufficiently low, the alternative technologies can compete without subsidies with oil, coal and natural gas. This argument for subsidies for alternative technologies is similar to the infant industry argument in trade theory whose effectiveness is subject to debate.[54,55] Taxpayer subsidies and government mandates may be economically inefficient if they merely act to stimulate innovation by guaranteeing a market for specific commodities, and not allowing a choice of technology based on economic cost. It is less contentious if the government gives subsidies for R&D and pilot-testing of renewable energy technologies where the externality benefits are more certain. At issue is whether renewable energy firms should receive subsidies and if so, how much subsidy should be given.[56]

---

[53] In addition to domestic subsidies, some renewable energy projects may also receive grants from international or foreign organisations and may also qualify for carbon credits. The economic basis for foreign grants or subsidies would be the positive externalities to other countries generated by renewable energy, for instance, in the mitigation of global climate change.

[54] The infant industry argument in trade is based on a dynamic theory of comparative advantage: temporary protection of domestic firms can generate positive learning and spillover effects. The difference though is that instead of a trade instrument (e.g. a tariff), we have a domestic instrument, a production subsidy for an alternative energy technology. If there exists dynamic production externalities, then the production subsidy, a domestic policy instrument, is a first-best policy. The effectiveness of the policy depends on whether the dynamic efficiency improvements outweigh the subsidy costs while its political economy aspect depends on whether the subsidy is time-bound and perceived to be temporary to motivate firms to innovate.

[55] The failure of import-substitution strategies, based on the infant industry argument, in some countries in Latin America and Africa and the much better economic performance of some Asian countries (e.g. Singapore, South Korea, Hong Kong and Taiwan) that pursued export-oriented strategies lend empirical evidence against infant industry protection.

[56] For instance, there are questions on whether Germany, which has relatively less sunny hours, should have invested as much in solar technology. Another example is that of Solyndra, a solar panel maker that received government subsidies in the US, which has declared bankruptcy. Solyndra received US$535 million in federal guaranteed loans. It has been argued that Solyndra is a victim of technological success, that is, Solyndra cannot compete with the solar panels from China whose prices have dropped due to economies of scale. Reports suggest that China provided around US$30 billion in subsidies to its solar industry in 2010 and currently accounts for 54 % of the world's solar panels. On the other hand, an argument supporting government subsidies is that federal loan programmes require the government to take risks that the private sector will not and, due to the very nature of risk-taking, there will be failures and successes involved. The question of whether renewable energy firms should receive subsidies persists; see Nocera, Joe, "The Phony Solyndra Scandal", in: *The New York Times* (23 September 2011), at: http://www.nytimes.com/2011/09/24/opinion/the-phony-solyndra-scandal.html?_r=1&pagewanted=print (7 January 2012).

Another view is that fossil fuels, in the absence of carbon or pollution taxes, are underpriced since their prices do not reflect the environmental damages generated by their use. As environmental economists would put it, ideally, the prices should reflect the cost of these externalities. In this scenario, alternatives to fossil fuel are at a disadvantage because fossil fuels are priced lower than they should be. This argument may offer a justification for subsidies for alternative energy as a second-best solution in the absence of emission taxes. The first-best approach is to impose pollution taxes on fossil fuels but this policy "creates visible costs and diffuse benefits, and politicians prefer the opposite: visible beneficiaries and diffuse costs to the taxpayers and the economy".[57]

4.6.4 Self-Sufficiency and Regional Cooperation

The question on whether energy self-sufficiency and energy independence is a desirable goal in pursuit of energy security is similar to the issue of food security and the notion of rice self-sufficiency in countries such as the Philippines. With a global economy and the interdependence of markets, energy self-sufficiency may not be a realistic objective and does not necessarily ensure energy security for a country. Energy autarky is generally not even a feasible solution if governments want reliable and affordable energy supplies to sustain economic growth. Just like trade in the usual goods and services, protectionism is likely to lead to inefficient resource allocation (you produce goods that others can provide more efficiently) and domestic consumers paying higher prices for energy.

In a global economy, countries benefit not only from access to global markets for goods and services but also from the flow of capital and technology across countries. In the Philippines, for example, the solar panels and wind turbines installed are imported. Similarly, base load power plants mostly use imported coal, given the technical limits on how much local coal can be substituted for the higher quality imported coal. Domestic energy resources may be efficiently used, as determined by the market, as long as it is cheaper to do so. The higher fossil fuel prices also provide incentives for the exploration and development of domestic energy sources, with investments by both local and foreign investors.

The goal of energy self-sufficiency in Southeast Asian countries seems to be a non-issue given the trade and investment relationships that exist among member countries. The terms, energy self-sufficiency and energy independence, have populist appeal, but are contrary to the ASEAN thrust of promoting free trade in

---

[57] van Doren, Peter, "The Evidence is Mixed", in: *The New York Times* (21 September 2011), at: http://www.nytimes.com/roomfordebate/2011/09/20/why-isnt-the-us-a-leader-in-green-technology/the-evidence-is-mixed (7 January 2012).

goods and services and letting comparative advantage determine the direction of production and trade.

The general direction for energy security for the ASEAN region, therefore, is towards greater regional cooperation and trade in energy in order to diversify away from Middle East sources of oil. Other options being considered by ASEAN are oil stockpiling and the use of nuclear power for electricity generation. Plans, such as the ASEAN Power Grid (APG) and the Trans ASEAN Gas Pipeline (TAGP), aim to expand cross-border connections for more diversified and efficient utilisation of resources. The ASEAN Petroleum Security Agreement (APSA)[58] seeks to enhance petroleum security and reduce emergency risks in the region. The APSA, among other initiatives, establishes a petroleum-sharing scheme to assist member states in emergencies arising from petroleum supply shortages and provides a platform for undertaking stockpiling initiatives among member states.[59]

4.7 Conclusion

Energy security continues to be a major concern for Southeast Asian countries in the face of increasing world oil prices, growing demand for energy due to faster economic growth and increased population, and greater environmental concern. This chapter looked at the energy security concerns of Southeast Asian countries and how they address these concerns, in particular the socioeconomic aspect of energy security, which involves ensuring reliable and adequate energy supply at affordable prices to support sustainable economic growth. In Southeast Asia, the energy security issue is coupled with environmental goals, as in most other countries, and with the social issue of access to modern energy services, especially in the lower-income countries.

The Southeast Asian countries subscribe to the core strategies of development of new sources, diversification of supply, and greater use of renewables; promotion of competitive markets; greater energy efficiency in use, particularly in transport and buildings; missionary electrification for remote and off-grid areas; and regional cooperation. The generally accepted social goals of economic policies—sustained economic growth and improved economic well-being—are often couched in the catchphrases: sustainable development and quality of life. The economic justifications for intervention by the government—assumed to have a longer time horizon—in economic activities are the presence of market imperfections and income redistribution. That is, the objectives of the

[58] ASCOPE, *ASEAN Petroleum Security Agreement (APSA)*, at: http://ascope.org/component/content/article/6-projects/30-apsa.html. (5 October 2011).

[59] Ho, Abigail L., "ASEAN Energy Ministers OK Pact Assuring Region's Oil Supply: Region's Producers to Offer 'Friendship Prices'", in: *Philippine Daily Inquirer* (11 August 2008), at: http://newsinfo.inquirer.net/inquirerheadlines/nation/view/20080811-153865/ASEAN-energy-ministers-OK-pact-assuring-regions-oil-supply (7 January 2012).

government are to ensure the smooth functioning of the economy or the markets (e.g. through regulation) and ensuring social cohesion and stability, its equity or redistribution function.

The efficiency of the energy markets can be greatly enhanced by correcting price signals through the reduction of fossil fuel subsidies, especially in energy resource-rich Brunei, Indonesia and Malaysia. On the supply side, exploration projects for oil, natural gas and coal in the Southeast Asian region continue and development of geothermal resources (e.g. in Indonesia and the Philippines) and hydropower resources, especially in the GMS, are ongoing. Singapore, resource-poor but an advanced economy, is investing in R&D of renewable energy technologies. Energy security for the Southeast Asian countries is promoted at the regional level by greater regional trade in energy goods and services, and cross-country investments, and through ASEAN initiatives, such as the proposed APG, TAGP and APSA.

Though the governments of ASEAN countries where fossil fuel subsidies are in place recognise the inefficiency of such subsidies and the need for their phase-out, they are constrained by political considerations and the potential social unrest that rising oil and other energy prices can generate. Fossil fuel subsidies distort relative prices making fossil fuel cheaper and inducing its greater consumption, and, as a means of helping the lower-income classes, has high leakage wherein more of the benefits go to the higher-income classes. Such price distortion also biases away from the development and use of renewable energy resources. Some countries, such as Indonesia and the Philippines, have moved in the direction of CCT schemes for income assistance to poorer households.

For developing ASEAN countries, the poverty issue in relation to energy markets remains a priority. With the ASEAN composed mostly of developing countries, the response of these countries has to take into account the well-being of the lower-income sectors of the population, as well as the capacity and willingness to pay of the general population for energy which also remain low relative to more advanced economies. Access to modern energy services—electricity and clean cooking facilities—remains a major concern in lower-income countries of the ASEAN. There is still wide scope for electrification, especially in rural areas, where renewable energy plays an important role especially for off-grid supply. The traditional use of biomass in cooking that is associated with negative health effects from indoor air pollution remains high in the lower-income countries.

Public acceptance of the use of renewable energy is also hampered by its current higher price compared to conventional energy sources. Technological developments in renewable energy that can lead to its large-scale commercialisation and lower price would have a significant impact in its use in the region. The question of whether renewable energy should be subsidised remains, but it is less contentious that government support for R&D of renewable energy technologies has positive social gains. The current high price of oil can also be viewed as providing an opportunity for greener energy and hence, if we do the transition to renewables and less polluting alternatives well, better environmental quality and a higher quality of life.

References

ADB (Asian Development Bank) (2011) Key indicators for Asia and the Pacific 2011, 42nd edn. Asian Development Bank, Manila

ASEAN (Association of Southeast Asian Nations) (2007) Cebu declaration on East Asian energy security, Cebu, Philippines, 15 Jan. ASEAN Secretariat, Jakarta. http://www.asean.org/23313.htm. 7 Jan 2012

ADB (Asian Development Bank) (2008) Reasons to be nervous: commodity prices and inflation. Asian development outlook 2008: workers in Asia. Asian Development Bank, Hong Kong, pp 28–36. http://www.adb.org/Documents/Books/ADO/2008/ado2008.pdf. 6 Jan 2012

ASEAN (Association of Southeast Asian Nations) (2007) Cebu declaration on East Asian energy security, Cebu, Philippines, 15 Jan. ASEAN Secretariat, Jakarta. http://www.asean.org/23313.htm. 7 Jan 2012

Elkind J (2010) Energy security: call for a broader agenda. In: Pascual C, Elkind J (eds) Energy security: economics, politics, strategies, and implications. Brookings Institution Press, Washington, DC, pp 119–148

Fiscal Policy Office, Ministry of Finance (Indonesia) (2011) Fuel subsidy policy in Indonesia. Presented at IISD conference on increasing the momentum of fossil-fuel subsidy reform: developments and opportunities. http://www.globalsubsidies.org/files/assets/ffs_gsiunepconf_sess2_askolani.pdf. 7 Jan 2012

Handayani SW, Burkley C (eds) (2010) Social assistance and conditional cash transfers: the proceedings of the regional workshop, July 2009. Asian Development Bank, The Philippines. http://www.adb.org/documents/conference/social-assistance-cct/proceedings-social-assistance-cct.pdf. 7 Jan 2012

Hogan WW (2009) Electricity market structure and infrastructure. In: Gallagher KS (ed) Acting in time on energy policy. Brookings Institution Press, Washington, DC, pp 128–161

IEA (International Energy Agency), 2009: Overview of energy trends in Southeast Asia: how will ASEAN energy needs be met? World energy outlook 2009. OECD/IEA, Paris

IEA (International Energy Agency) (2010) World energy outlook 2010. OECD/IEA, Paris

Jha S, Quising P, Camingue S (2009) Macroeconomic uncertainties, oil subsidies, and fiscal sustainability in Asia, ADB economics working paper series No. 150 Asian Development Bank, Manila. http://www.adb.org/Documents/Working-Papers/2009/Economics-WP150.pdf. 23 Feb 2012

Tay SSC, Fang N (2011) Examining nuclear energy ambitions in South-East Asia, August, SIIA special report 11/02, Singapore Institute of International Affairs [SIIA]

UNDP (United Nations Development Programme) (2007) Human development report 2007/2008: fighting climate change: human solidarity in a divided world. UNDP, New York

UNDP (United Nations Development Programme) (2010) Human development report 2010: the real wealth of nations: pathways to human development. UNDP, New York

World Bank (2011) World development indicators 2011. The International Bank, Washington, DC

Abbreviations

ADB	Asian Development Bank
APG	ASEAN Power Grid
APSA	ASEAN Petroleum Security Agreement
ASCOPE	ASEAN Council on Petroleum
ASEAN	Association of Southeast Asian Nations
Btu	British thermal unit

CCT	Conditional cash transfer
CNG	Compressed natural gas
CSPS	Centre for Strategic and Policy Studies
DSWD	Department of Social Welfare and Development, Republic of the Philippines
FIT	Feed-in-tariffs
GDP	Gross domestic product
GHG	Greenhouse gas
GMS	Greater Mekong Subregion
GST	Government service tax
HDB	Housing and Development Board
IMF	International Monetary Fund
kWh	Kilowatts hour
LIHD	Low-input high-diversity
LPG	Liquefied petroleum gas
MW	Megawatts
NCR	National Capital Region
NPC	National Power Corporation
NREB	National Renewable Energy Board
NREL	National Renewable Energy Laboratory
OECD	Organisation for Economic Co-operation and Development
PPP	Purchasing power parity
R&D	Research and development
SIIA	Singapore Institute of International Affairs
SPUG	Small Power Utilities Group
TAGP	Trans ASEAN Gas Pipeline
toe	Tons of oil equivalent
UNDP	United Nations Development Programme
UNESCAP	United Nations Economic and Social Commission for Asia and the Pacific
US	United States
U-Save	Utilities-Save
VAT	Value-added tax

Author Biography

Maria Nimfa F. Mendoza (Singapore): Assistant professor at the School of Economics, University of the Philippines (Diliman), The Philippines. She has a BS in Applied Mathematics and MS in Statistics from the University of the Philippines at Los Baños and a PhD in Economics from the University of British Columbia. Her fields of interest are public economics, environmental economics and energy economics. She also teaches mathematical economics and econometrics.

Chapter 5
An Environmental Perspective on Energy Development in Indonesia

Fitrian Ardiansyah, Neil Gunningham and Peter Drahos

Abstract Indonesia faces an energy trilemma on the energy security, climate change goals and energy poverty fronts. Policies that focus exclusively on one prong of the trilemma may lead to unacceptable consequences in the others. Conceiving the predicament as a trilemma will encourage a more unified approach to its problem solving. Successful management will require a search for policy complementarities—the likeliest source of which may be the renewable energy sector—that allow the country to move forward on all three fronts. A reform of its bureaucracy to address implementation gaps in its energy policy will also be needed. The reduction in transaction costs associated with the implementation of Indonesia's energy policy could be used as a broad criterion when considering these necessary changes.

F. Ardiansyah (✉)
WEH Stanner Room #1.38, Crawford School of Economics and Government,
The Australian National University (ANU), 132 Lennox Cross, Canberra,
ACT 0200, Australia
e-mail: fitrian.ardiansyah@anu.edu.au
http://fitrianardiansyah.wordpress.com

N. Gunningham
National Research Centre for Occupational Health and Safety Regulation,
RegNet, Research School of Pacific and Asian Studies,
The Australian National University (ANU), Canberra,
ACT 0200, Australia
e-mail: neil.gunningham@anu.edu.au
http://www.anu.edu.au/fellows/ngunningham/

P. Drahos
Centre for Governance of Knowledge and Development,
RegNet, Research School of Pacific and Asian Studies,
The Australian National University (ANU),
Canberra, ACT 0200, Australia
e-mail: peter.drahos@anu.edu.au
http://www.anu.edu.au/fellows/pdrahos/

M. Caballero-Anthony et al. (eds.), *Energy and Non-Traditional Security (NTS) in Asia*, SpringerBriefs in Environment Security, Development and Peace 1,
DOI: 10.1007/978-3-642-29706-9_5,

Keywords Energy security · Energy poverty · Renewable energy · Biofuels · Climate change · Environmental impacts · Deforestation

5.1 Introduction

Indonesia[1] (Fig. 5.1) is faced with a policy trilemma consisting of three prongs: energy security, climate change mitigation and energy poverty reduction. Simply put, achieving one of these policy goals will entail compromise on the achievements of at least one of the others. For example, Indonesia's most probable path to energy security is to heavily rely on its coal and gas reserves. However, its President, Susilo Bambang Yudhoyono, has made commitments under the Copenhagen Accord to reduce greenhouse gas (GHG) emissions by 26 % by 2020 and to increase the use of renewable energy so that it accounts for 25 % of total energy production by 2025. Any energy security pathway that is based on fossil fuels will make achieving such climate mitigation targets difficult. Indeed, increased emissions from fossil fuel combustion are projected to increase GHG emissions in Indonesia fourfold by 2030.[2]

Critics might argue that the above trilemma is overstated, given that it is possible to identify policy complementarities as well as conflicts. For example, Indonesia has more geothermal energy potential than any other country—27,000 MW of potential geothermal energy reserves[3]—that if exploited properly would go a long way to help Indonesia meet its energy security and climate mitigation goals. However, such development, while desirable, is fraught with hurdles. In a politically frictionless world, Indonesia might have been able to transit seamlessly to geothermal energy; in reality though, Indonesia has to make this transition in keeping with present institutional structures and political networks under which the price of fossil fuel is not only heavily subsidised but also centrally set. That the state electricity generator, Perusahaan Listrik Negara (PLN),[4] which purchases all power, has set the price of energy from renewable sources such as geothermal at a rate so high as to make its production untenable does not help either. The commercialisation of geothermal energy, in some

[1] The Government of Indonesia, "The Geography of Indonesia", at: http://www.indonesia.go.id/en/indonesia-glance/geography-indonesia.html (29 June 2011).

[2] Fiscal Policy Office (Ministry of Finance); World Bank, "Phase 1 Assessment of Emissions—Key Findings", in: *Low Carbon Development Options for Indonesia*, at: http://www.esmap.org/esmap/sites/esmap.org/files/factsheet04.pdf (16 March 2011).

[3] See Allard, Tom, "Indonesia Goes to Ground for Energy", in: *Sydney Morning Herald* (1 May 2010), at: http://www.smh.com.au/environment/energy-smart/indonesia-goes-to-ground-for-energy-20100430-tzbv.html (16 March 2011).

[4] For PLN's full profile, see "Company Profile", at: http://www.pln.co.id/eng/?p=102 (12 August 2011).

Fig. 5.1 Map of Indonesia. *Source* CIA 2011 (CIA, "East and Southeast Asia: Indonesia", in: *The World Factbook* (14 June 2011), at: https://www.cia.gov/library/publications/the-world-factbook/maps/maptemplate_id.html (29 June 2011)

circumstances, also involves a direct conflict with the long-established, well-organised and powerful forestry industry, as an estimated 60 % of geothermal energy sources are located in forest areas that are also subject to the recently enacted laws on pristine forests, including stricter conditions under which licences are to be issued (Girianna 2009, p. 2). Compounding the problem is the fact that nearly half of these resources belong to local governments while the rest is in the hands of state-owned companies.[5]

One message this chapter drives home is that the renewable energy targets that Indonesia is setting for itself risk becoming an unattainable wish list if they are not accompanied by a mix of regulatory strategies that are sensitive to the feasibility constraints being imposed by Indonesia's institutions and politics. For example, the fuel subsidy that was used during the Suharto regime to ensure the availability of cheap energy has turned into a huge fiscal burden for the state, amounting to nearly 21 % of total government expenditure in 2005, that has hitherto continued to rise (Resosudarmo et al. 2010, p. 2). Addressing such feasibility constraints and others like it will require Indonesian policymakers to chart regulatory and governance courses that accompany the political goals and promises being made by the current Indonesian President. Then again, as is generally the case, it is easier to provide a thick description of problems and suggest general solutions, such as increasing investments in renewable energy, than it is to show how one might arrive at these solutions in a world where national political networks have to devise and execute the 'how to' prescriptions as well as bear the resulting costs.

[5] "Potret Keadaan Hutan Indonesia Periode 2000–2009" (The Picture of Indonesia's Forests 2000-2009), in: *FWI (Forest Watch Indonesia)* (27 July 2011), at: http://fwi.or.id/?p=306 (31 August 2011).

The rest of the chapter sets out the tensions and trade-offs that characterise the prongs of Indonesia's trilemma. Climate change, as will be elaborated later in this chapter, is a proxy for a set of nested environmental public good problems that are linked in various ways to available energy options in Indonesia. The chapter will explore these options, review both the obstacles and opportunities for the development of renewable, clean and sustainable energy, and assess whether these can provide effective solutions to Indonesia's energy challenge, particularly in terms of resolving its energy trilemma. Prior to this, however, the chapter will provide a profile of energy development in the country and its associated environmental impacts. This discussion seeks to identify a balanced solution that addresses both environmental protection and economic development when it comes to securing current and future energy for the country. The chapter will conclude by summarising its findings and discussing key possible steps towards the resolution of the energy trilemma. The final section suggests some areas where Indonesia needs to develop a 'how to' regulatory strategy in order to achieve the goals it has set for itself on energy diversification and climate change mitigation. In particular, the authors advocate the development of strategies that lower the transaction costs of Indonesia's decentralised governance arrangements around energy investment and regulation.

The analysis presented in this chapter was supported by literature and data reviews as well as semistructured interviews and focus group discussions with key stakeholders. The latter included officials from state agencies, non-state actors (including non-governmental organisations), international organisations, and Indonesia specialists situated both in Indonesia and other countries. Overall, 22 interviews were conducted,[6] with a focus on contemporary policies, politics and the institutional construction of the energy sector, thereby providing a basis for the critical assessment of the environmental aspect of energy development in Indonesia.

5.2 Energy Mix: Context and Challenges

Indonesia is the largest energy producer and consumer in Southeast Asia. Based on the levels of energy elasticity and intensity, Indonesia falls into the category of 'consumptive' or 'close to wasteful'. For example, the energy elasticity[7] figures for Indonesia (1.04–1.35) for the period between 1985 and 2000 (Koalisi Energi 2005, p. 3) are relatively high when compared with corresponding figures of

---

[6] WWF-Indonesia commenced a climate and energy programme in 2003 whose work, until the writing of this chapter in 2011, has increasingly gained recognition both nationally and internationally. The first author served as a director in this programme between 2007 and 2010. Between late 2010 and early 2011, he served as advisor to WWF-Indonesia and was involved in the energy governance research of Indonesia that was led by Professors Neil Gunningham and Peter Drahos of the ANU.

[7] Energy elasticity is defined as the percentage change in energy consumption to achieve 1 % change in national GDP.

developed countries (0.55–0.65) for the same period. Similarly, the energy intensity of Indonesia (index, 500) is five times that of Japan (index, 100) and higher than that of North America (index, ~300), Organisation for Economic Co-operation and Development (OECD) countries (index, ~200), and even Thailand (index, ~350) (Koalisi Energi 2005, p. 3). Historical data shows that demand increases in line with, if not faster than, economic development and population growth (ESDM 2010, pp. 2–3). From 1990 to 2006, for instance, Indonesia's compound annual growth rate of total final energy consumption (excluding biomass) was about 5.5 %, or an increase from 245 millions of barrels of oil equivalent (mmboe) to 577 mmboe.[8] According to the Green Policy paper released by its finance ministry, total energy demand is growing by around 7 % annually, as the transport and industrial sectors grow and as households become more affluent (Ministry of Finance 2009, p. 5).

Indonesia appears to be struggling to keep up with its energy demands, and its dependency on oil imports to fuel its economy and cope with its problems of energy security seem no longer economically viable. Soaring average global oil prices, reaching US$113.61 per barrel in June 2011,[9] have placed considerable strains on the Indonesian economy. The state budget is burdened by heavy oil and electricity subsidies[10]—subsidy costs to the government were estimated at US$9.78 billion in 2010[11] and US$3.68 billion as of March 2011.[12] The high price of oil on the global market has also caused a spike in electricity generation costs upstream. The transport sector, much like the energy sector, is a large consumer of fossil fuels too—consumption increased from 6 million tonnes of oil equivalent (mtoe) in 1980 to 25 mtoe

---

[8] Ardiansyah, Fitrian; Wardhini, Indra Sari; Suhud, Muhammad, "Road to Copenhagen: Helping Indonesia Advance as its Emissions Decrease", in: *The Jakarta Post* (27 October 2009), at: http://www.thejakartapost.com/news/2009/10/27/road-copenhagen-helping-indonesia-advance-its-emissions-decrease.html (16 January 2012).

[9] "World Crude Oil Prices", in: *US-EIA (U.S. Energy Information Administration)* (17 June 2011), at: http://www.eia.gov/dnav/pet/pet_pri_wco_k_w.htm (28 June 2011).

[10] Under the Suharto regime, which was anxious to maintain social stability, there was an emphasis not only on food but also energy security. As in a number of other developing countries, the central government subsidised the price of a variety of energy products, including low-octane gasoline, kerosene, diesel, liquefied petroleum gas (LPG) and electricity, with the aim of ensuring that energy was cheap and available (Agustina et al. 2008, p. 12). As long as the price of oil was low and the value of the rupiah relatively high, the subsidies remained modest. However, the Asian financial crisis of 1997 resulted in a substantial fall in the value of the rupiah even as the price of crude oil rose substantially in the early 2000s. Fuel subsidies increased markedly from 1998 to 2000 following the sharp depreciation of the rupiah relative to the US dollar, peaking in 2000 and accounting for 28.6 per cent of total spending (Agustina et al. 2008, p. 12). In 2008, fuel and electricity subsidies together reached US$20.5 billion (Agustina et al. 2008, p. 3).

[11] Suharmoko, Aditya. "Govt Likely to Increase Fuel Subsidy: Minister", in: *The Jakarta Post* (7 August 2010), at: http://www.thejakartapost.com/news/2010/07/08/govt-likely-increase-fuel-subsidy-minister.html (16 March 2011).

[12] Kertiyasa, Martin Bagya, "Maret, Subsidi Energi Rp32,3 Trilyun" (March, Energy Subsidy Rp32.3 Trillion), in: *Okezone* (14 April 2011), at: http://economy.okezone.com/read/2011/04/14/20/445920/ (15 April 2011).

in 2005 or 48 % of the national fuel consumption (Timilsina and Shrestha 2009, p. 4528). The demand for energy from this sector is projected to increase rapidly,[13] and this growing demand coupled with the high cost of petroleum has prompted Indonesia to urgently seek fossil fuel substitutes, including vegetable oils.

A large amount of this demand has been met by fossil fuels, mainly oil. Since 2004, the country is a net importer of both crude oil and refined products (Sa'ad 2009, p. 4391). In 2004, for instance, production averaged 1,100 tbd while consumption hit 1,200 tbd (Sa'ad 2009, p. 4392). With a production capacity of 0.5 billion barrels per year and increasingly limited oil reserves, it is estimated that Indonesia's remaining 10 billion barrels of oil reserves will be exhausted in less than 20 years (Koalisi Energi 2005, p. 3). Should no new reserves be found, given its increasing demands for energy and a 'business as usual' approach, Indonesia is projected to become a significant oil-importing country in less than two decades.

While its oil reserves may be diminished, the country remains extremely rich when it comes to energy resources—Indonesia is widely known to have bountiful natural resources, including the renewable forms of energy such as run-of-the-river hydropower, geothermal, solar and biomass. It is the second largest thermal coal exporter after Australia[14] and remains a major liquefied natural gas (LNG) exporter (IEA 2008, p. 109). It has the world's tenth largest reserves of natural gas accounting for some 40 % of the world's geothermal reserves (Leitmann et al. 2009, p. 67), which albeit underutilised might potentially meet some 40 % of its electricity demands (Leitmann et al. 2009, p. 67).[15] Biofuels are being explored as well, including those sourced from palm oil, *Jathropha curcas* and castor oil.

The development of renewable sources over the past several years has, however, progressed very slowly. According to a 2009 World Bank report, for example, these resources were still largely unexploited and underdeveloped despite their high potential as clean domestic energy sources (World Bank 2009, p. 2). Domestic power generation has remained a particular challenge, with the majority of energy sources for such generation currently being conventional thermal sources that include fossil fuels, such as oil, coal and natural gas—less than 20 % of power generation is accounted for by hydroelectric, geothermal and other renewable sources.[16] Compounding matters further, Indonesia has

---

[13] "Indonesia Overtakes Thailand in Car Sales", in: *The Independent* (1 August 2010), at: http://www.independent.co.uk/life-style/motoring/indonesia-overtakes-thailand-in-car-sales-2040988.html (16 March 2011).

[14] "Indonesia May Ban Low Quality Exports from 2014", in: *Bloomberg* (24 January 2011), at: http://www.steelguru.com/raw_material_news/Indonesia_may_ban_low_quality_coal_exports_from_2014/187736.html (16 March 2011).

[15] President Susilo Bambang Yudhoyono has announced plans for Indonesia to become the world's leading geothermal nation, with 44 plants to be built by 2014 and its 4,000 MW capacity rising to 9,000 MW by 2025; see Allard, Tom, "Indonesia Goes to Ground for Energy", in: *Sydney Morning Herald* (1 May 2010), at: http://www.smh.com.au/environment/energy-smart/indonesia-goes-to-ground-for-energy-20100430-tzbv.html (16 March 2011).

[16] "Share of Total Primary Energy Supply in 2008: Indonesia", in: *IEA (International Energy Agency)* (2010), at: http://www.iea.org/stats/pdf_graphs/IDTPESPI.pdf (16 March 2011).

complicated its energy security over the last 20 years by selling a substantial part of its oil and coal (and to some extent, gas) to a number of energy-hungry countries under long-term contracts.[17] Steps are being taken to put this right—Indonesia's energy policy has in recent times seen a sea change, as the country's political leaders have come to appreciate the serious difficulties that are to be faced in securing and providing basic energy to its citizens. For instance, Indonesia is a substantial exporter of LNG, ranking third after Qatar and Malaysia, and a substantial percentage of its energy resources are exported. In 2004, 70 % of Indonesia's LNG exports went to Japan, 20 % to South Korea and 10 % to Taiwan.[18] However, new deals between Japanese companies and Indonesia's PERTAMINA (Perusahaan Tambang dan Minyak Negara)[19] over LNG are for considerably less volumes than previously considered, in line with a general decrease in Indonesia's LNG exports following the realisation of decreasing supply amid increasing domestic demand.[20]

Indonesia has plans to promote coal and biofuels for addressing energy security, but these may create additional problems in terms of its climate change mitigation targets. The use of coal entails emission costs just as aggressive biofuel development may contribute to GHG emissions. According to a recent report, although presently more than 90 % of Indonesia's GHG emissions are from the forestry and land-use sectors, any rapid increase in emissions from the energy sector in the future—should more coal power plants be built and biofuel crops replace the remaining forests and peatlands—would make Indonesia an important contributor to GHG emissions (USAID 2008, p. 16). Correspondingly, reports also project raised carbon dioxide (CO_2) emissions by 2030 (DNPI 2010, p. 25). Such increase in emissions will exacerbate climate change to which Indonesia, an archipelagic nation, is particularly vulnerable (Ministry of Environment 2009, p. 12). Indeed, climate change is credited with creating an increasing threat to economic activities within the country, including damage to assets from extreme weather, loss of agricultural outputs due to droughts, flooding, transboundary haze and unseasonal

---

[17] As to China's resource relationship with Indonesia, see Chaney, Joseph; Wee, Sui-Le, "Indonesia's Resources Appeal Grows on China", in: *Reuters* (17 February 2010), at: http://www.reuters.com/article/2010/02/17/us-dealtalk-indonesia-resources-idUSTRE61G2K220100217 (16 March 2011); for a general, but somewhat dated energy profile, see "Indonesia Energy", at: http://wn.com/s/indonesiaenergy_old1/ 16 March 2011.

[18] "Indonesia", in; *Indonesia Energy Info* (July 2004), at: http://wn.com/s/indonesiaenergy_old1/ (16 March 2011); Alfian; Fox, David, "Indonesia LNG Exports to Fall to 362 Cargoes This Year", in: *Reuters* (18 January 2011), at: http://www.reuters.com/article/idUSL3E7CI0UK20110118 (16 March 2011).

[19] PERTAMINA is Indonesia's state-owned oil and gas company. For the company's full profile, see "Company Profile", at: http://www.pertamina.com/index.php/home/read/company_profile (12 August 2011).

[20] "Raft of New Deals as Japan and Indonesia Agree LNG Terms", in: *Business Monitor International* (March 2008), at: http://www.oilandgasinsight.com/file/63096/raft-of-new-deals-as-japan-and-indonesia-agree-lng-terms.html (16 March 2011).

weather, and disruption to business from infrastructure damage or disruption (Ministry of Environment 2009, p. 12).

Last but not the least, the policies that Indonesia chooses vis-à-vis climate change mitigation and energy security are bound to have a profound impact on the final component of the energy trilemma, that is, energy poverty. Fuel subsidies, as is the case in many developing countries, play a pivotal role in Indonesia's politics. As a G20 member, Indonesia has agreed to phase out fuel subsidies, a brave promise given that some 50 % of its 225 million inhabitants exist on less than US$2 a day. Some 80 million people in Indonesia, primarily those in rural areas, still lack access to electricity[21] and many others experience energy insecurity in the form of frequent blackouts and brownouts. For those without electricity and who subsist on less than US$2 a day, traditional biomass (such as the three-stone fire) is the principal source of energy; it is known to cause several problems, including premature death from household air pollution—more poor people die prematurely from this than do from malaria or tuberculosis (OECD/IEA 2010, p. 13). Certainly, access to electricity for poor people is an obvious priority as is access to clean cooking fuel.

Fuel subsidies do a poor job of benefiting the rural poor, tending to favour urban populations through the provision of cheap electricity (Resosudarmo et al. 2010, pp. 14–15). Nonetheless, while it is true that the rich benefit disproportionately from the subsidy (Pallone 2009, p. 5), it is also true that the poor receive some immediate benefit as well. Although removing fuel subsidies would send the correct price signals to Indonesia's energy markets, it would in the short term lessen access to energy for poor people and deepen the country's energy poverty. Driving the poor into further energy poverty may also drive out hope for some, a dangerous path for any government to tread. For this reason, simply sweeping the subsidy away would be full of political risk.

There is uncertainty as to whether the Indonesian political elite will act quickly to dispose off fuel subsidies. As one official the authors interviewed pointed out: "for a developing country, the subsidy is not just about economic policy; it's very dependent on political policy. Our proposal to make an adjustment to the tariff is restricted by Parliament, so we can't do it. It's not easy! So you shouldn't push developing countries too far". This was reiterated by another interviewee from the Ministry of Finance who pointed out that while the ministry did have plans to phase out subsidies, one had "to remember, if we remove the energy subsidy, we get higher inflation and higher poverty …". There are no obvious easy resolutions to such problems, and it is, therefore, not surprising that Indonesian policymakers are inclined to talk in vague terms about incremental changes to be phased in at some future point in time comfortably beyond the current political cycle.

---

[21] "Snapshot of Electrification throughout ASEAN", in: *Asian Trends Monitoring* (29 July 2010), at: http://www.asiantrendsmonitoring.com/2010/07/29/snapshot-of-electrification-throughout-asean (16 March 2011); Barbotte, Daphné, "Country Spotlight: Indonesia", in: *REEEP* (5 September 2010), at: http://www.reeep.org/index.php?id=443&special=showHotTopic&iHotId=861&sQuiteName=news&iQuiteId=436 (16 March 2011).

Nevertheless, over time, one can expect the existing subsidies to be removed as their costs mount and more targeted ways are found to compensate poor people. Senior Indonesian bureaucrats are cognisant that subsidies are a problem that will have to be dealt with using long-term negotiations and not sweeping radical reform. There is not much else that can be said about this issue.

5.3 Energy Development: Options, Impacts and Failings

The pressures on energy security and the economy have prompted Indonesia to reconfigure its energy policy. With supplies of domestic oil dwindling, reliance on oil imports increasing, and energy demand and the need to combat energy poverty growing, Indonesia is looking, with increasing urgency, for additional conventional sources of power and alternative sources of energy. The cheapest short-term source of electricity for Indonesia is definitely coal—PLN has plans, with regard to electricity, to develop and promote coal-dominated power plants for Java and diesel-dominated plants for the outer islands. Options that can help Indonesia meet its demand for energy while ensuring that negative environmental impacts are avoided or reduced, are also being pursued. Indonesia's huge geothermal reserves if utilised correctly, would reduce its GHG emissions and help it gain carbon credits. Similarly, biofuel development, if undertaken responsibly for non-forested and non-peatland areas, would contribute positively to climate change mitigation and economic development.

5.3.1 Coal

Coal is central to Indonesia's energy mix and policies, both present and future, and emerges as an obvious choice in Indonesia's hunt for a suitable energy source—the country has a plentiful supply of the resource and coal-fired power stations use well-established technology that can be purchased relatively cheaply and assembled relatively quickly. These cost factors led, in 2006, to a government 'crash programme' that aimed to develop 10,000 MW from coal-fired plants that would be readied in the coming years, as stipulated in the Presidential Decree No. 71 of 2006 (Leitmann et al. 2009, p. 27). Under the programme, most plants were purchased cheaply from China and quickly assembled. While the programme did not provide long-term efficiency (interview data suggests that the plants were cheap because they were dirty and inefficient), the plants did bring about short-term improvement in the operation of electricity grids. The programme was also attractive because it shifted energy generation away from the increasingly expensive and imported oil, which is slated for gradual phase-out.

Following the first programme, a second-phase crash programme, to be realised between 2009 and 2018, has also been introduced with the same projected increase in total intended capacity and at an estimated cost of US$21.3 billion.[22] While interviewees consistently suggested that even the second phase was primarily about coal, formal statements of Indonesian officials were to the contrary, suggesting that 60 % of the new capacity would come from renewable resources, and that, in particular, approximately 5,000 MW or 48 % would be from geothermal resources and 12 % from hydro resources (Girianna 2009, p. 1).

Be that as it may, a dramatic surge in coal-fired plants will greatly increase GHG emissions and contribute to climate change. Notwithstanding current trends toward renewable sources for the energy and transport sectors, Indonesia's fossil fuel-based emissions are forecasted to triple by 2030 (World Bank 2009, p. 1). In the case of coal, per capita GHG emissions from fossil fuel consumption in Indonesia grew faster than in China and India even for the years between 1999 and 2004 (World Bank 2009, p. 2). With the government further planning to significantly increase the share of coal in the country's power generation fuel mix, GHG emissions are anticipated to grow faster than the economy (World Bank 2009, p. 2). An analysis from the Dewan Nasional Perubahan Iklim (DNPI; also known as National Council on Climate Change [NCCC])[23] suggests that the country's emissions from the power sector are expected to grow sevenfold from 110 metric tonnes (tons) of CO_2 equivalents ($MtCO_2e$) in 2005 to 810 $MtCO_2e$ in 2030 due to strong demand growth and an increasing dependence on coal (DNPI 2010, p. 25). DNPI projects that emissions from the power sector in 2030 will exceed those from the peat sector today (DNPI 2010, p. 25).

The ill effects of the use of coal as the predominant source of energy, however, do not end with increased emissions. Among others, coal may lead to significant environmental consequences, such as smog and acid rain,[24] given the fact that gases are given off and ash particles released whenever coal is burnt. The sulphur in coal combines with oxygen to form sulphur dioxide, which can be a major source of air pollution if emitted in large enough quantities.[25] When coupled with other industrial pollutants and motor vehicle emissions, the negative impact of coal on the health of urban populations can be serious, especially in terms of respiratory diseases (Leitmann et al. 2009, p. 65).

---

[22] ESDM, "Second Phase of Electricity Crash Program and Development of Transmission and Distribution Networks", in: *The Free Library* (2009), at: http://www.thefreelibrary.com/Second+phase+of+electricity+crash+program+and+development+of...-a0198850375 (16 March 2011).

[23] Dewan Nasional Perubahan Iklim (DNPI) or National Council on Climate Change (NCCC) is an institution set up by President Susilo Bambang Yudhoyono in July 2008 to coordinate climate change-related activities within Indonesia. The council is specifically tasked with the role of convening different stakeholders in Indonesia to create consensus around the opportunities and challenges related to climate change.

[24] Ardiansyah, Fitrian, "Indonesia's Energy Dilemma", in: *The Jakarta Post* (7 July 2010), at: http://www.thejakartapost.com/news/2010/07/06/climate-solutions-indonesia's-energy-dilemma.html (16 March 2011).

[25] National Energy Foundation, "Background Information: How Does Burning Coal Affect the Environment", at: http://www.coaleducation.org/lessons/twe/envi.htm (29 June 2011).

5.3.2 Palm Oil

Indonesia has expanded its capacity for palm oil production and is now a leading producer in world terms. Like other vegetable oils, palm oil can be used as a fuel in vehicles and for electricity and heat generation. The Government of Indonesia, therefore, has promoted a plan to use oil palm-based biodiesel to reduce its dependency on fossil fuels by 10 % in 2010 and 20 % by 2025 (Ariati 2007, p. 12).[26] With incentives, such as the allocation of US$1.15 billion from the state budget for infrastructure related to biofuel development, tax incentives (30 % reductions on net profits)[27] and interest rate subsidies (10 %) for smallholders,[28] it is projected that numerous producers will take advantage of this favourable economic environment and rapidly expand palm oil production.

The development of biofuel crops, particularly palm oil, could further increase GHG emissions if palm oil plantations replace forests and peatlands. For example, in the Indonesian provinces of North Sumatra and Bengkulu, and in Peninsular Malaysia, 38, 35 and 27 %, respectively, of peat-swamp forest were converted to oil palm plantations by the early 2000s—this conversion led to the release of about 140 million tons of carbon from biomass above ground and 4.6 million tons of carbon from peat oxidation below ground.[29] According to Koh and Wilcove's estimates (2008, p. 62), the forest area converted to oil palm plantations in Indonesia was 1.7 million ha between 1990 and 2005,[30] meaning that over 56 % of oil palm expansion occurred at the expense of natural forest cover for the period. In terms of peatlands, Hooijer et al. (2006, p. 10) state that of the 10.3 million ha of land allocated to palm oil concessions, 2.8 million ha ($\sim$27 %) are located on peatlands. According to the Badan Perencanaan Pembangunan Nasional (BAPPENAS; also known as National Development Planning Agency) (BAPPENAS 2009, p. 5), as of 2006, plantation licenses (i.e. predominantly for oil palm) on peatlands totalled 1.3 million ha.

The environmental problems that such conversions cause are exacerbated where drainage has been intensified to establish large-scale plantations. This results in very high annual emissions of up to a hundred tons of CO_2 per ha, compared to only a few tons from recently logged areas (Hooijer et al. 2006, p. 10, 12, 15, 17–33). Considerably more carbon may be released into the atmosphere when fire is used to clear forests and peatlands and make way for oil palm plantations.

[26] See Presidential Instruction No. 5 of 2006 on National Energy Policy.

[27] See Government Regulation No. 1 of 2007 on Income Tax Facilities to Investment Activities in Specific Industries and Particular Regions.

[28] See Ministry of Forestry Regulation No. 117/PMK.06/2006 on the Credit for the Development of Biofuel Energy and Plantation Revitalization.

[29] Vaidyanathan, Gayathri, "Counting the Carbon Cost of Peatland Conversion", in: *Nature* (7 March 2011), at: http://www.nature.com/news/2011/110307/full/news.2011.139.html>?s=news_rss (29 June 2011).

[30] In their analysis, Koh and Wilcove (2008, p. 62) assessed that between 1990 and 2005, oil palm cultivated areas in Indonesia increased by 3 million ha.

Apart from forest and peat destruction for oil palm plantations causing GHG emissions, forest and peatland conversion to oil palm has led to the fragmentation of *high conservation value forests* (HCVF) and had other environmental (e.g. human-wildlife conflicts) and social impacts (e.g. land tenure conflicts) as well. There is therefore a need to ensure that using palm oil as a source for energy does not affect its availability as a food for the poor, who are least able to afford alternative sources (Colbran and Eide 2008, pp. 4–10), or deprive them of food that would otherwise be available if land were used differently. For example, the increasing involvement of conglomerates in the bioenergy business has led to fears of the marginalisation of small producers. There are also concerns about the insecurity of land ownership and tenure rights for rural communities that depend upon access to forest and agricultural resources and ecosystem services. Finally, Wilcove and Koh (2010, p. 999) argue convincingly that oil palm development and the destruction of forests pose the greatest immediate threat to biodiversity and forests in Southeast Asia.

Although contentious, there is no doubt that palm oil is increasingly being seen as a major renewable energy source. Where Indonesia is concerned, the demand for palm and other vegetable oils, in very large part, comes from the world market and not the domestic one. Certainly, it is the recent increase in global demand for palm oil as a biofuel source that has prompted such an expansion of the Indonesian palm oil industry. The European market, driven by the high cost of petroleum and political pressure from the climate change mitigation movement, is in particular importing ever-increasing amounts of palm oil, although as indicated below this trend may not continue so. The EU has set a target of 5.75 % of energy from renewable sources by 2012, with biofuels playing a major role (Tindale 2009, p. 2). The US has pledged to increase the use of renewable fuels from 7.5 billion gal in 2012 to 35 billion gal in 2017 (EPRINC 2007, p. 4). Already, as of 2005, around 1.5 million tons of palm oil were used for this purpose in power stations throughout Europe (Reinhardt et al. 2007, p. 12).

Opportunities in Europe for Indonesian palm biodiesel producers, however, are likely to be closing (Sheil et al. 2009, p. 18). A 2008 directive issued by the European Parliament on biofuels and renewable energy sources has proposed three criteria for acceptable biofuels: (a) land with high carbon stocks should not be converted for biofuel production; (b) land with high biodiversity should not be converted for biofuel production; and (c) biofuels should achieve a minimum level (35 %) of GHG savings (Sheil et al. 2009, p. 18). The future for palm biodiesel is therefore likely to lie increasingly within Indonesia itself (as is also the case with Malaysia), and perhaps in other key consumer countries outside the EU, such as China and India. It is against this backdrop that the world's palm oil production is expected to nearly double by 2020 (Teoh 2002, p. 9).

5.3.3 Geothermal Energy

Indonesia's geothermal resources have remained underutilised as pricing delays between the PLN and the government have marred its geothermal options. Although geothermal licences were granted in 2010, developers were unable to begin exploration because power-purchasing agreements with the PLN remained unsigned as it waited for the government's pricing approvals.[31] These snags were apparently resolved on 1 January 2011 when the PLN introduced an 18 % tariff hike ceiling, in line with the Energy and Mineral Resources Minister's Regulation No. 7 of 2010, as a temporary measure to resolve tariff discrepancies caused by the ministerial regulation in 2010. However, the cost of developing geothermal technology is still prohibitive, costing an estimated US$800 million for a 333 MW power plant, and completely beyond the capacity of the Indonesian government. Recognising the high risks of investment in geothermal resources, the World Bank announced a US$400 million commitment from their Clean Technology Fund in early 2010.[32] The project is aimed at doubling Indonesia's geothermal energy capacity.[33] In addition, the Lahendong Geothermal Plant is being financed by the World Bank, Asian Development Bank (ADB)[34] and Japan International Cooperation Agency (JICA).[35] JICA also undertook a study in March 2008 of the fiscal incentives that could accelerate Indonesia's geothermal capacity. The study

[31] Exploration rights were granted as follows: the Jaboi field in Aceh was awarded to a consortium led by PT Bukaka Teknik Utama; the Sorik Marapi field in North Sumatera to a consortium of Tata Power and Origin Energy; the Muara Laboh field in West Sumatera and the Gunung Rajabasa field in Lampung to PT Supreme Energy; the Jailolo field in Halmahera to Star Energy; the Sokoria field in Flores Island to Bakrie Power; the Tangkuban Parahu field in West Java to PT Indonesia Power; the Cisolok field in West Java to PT Rekayasa Industri; the Tampomas field in West Java to PT Wijaya Karya; and the Ungaran field in Central Java to PT Golden Spike Energy Indonesia; see Alfian, "Indonesia Geothermal Program Hung Up on PLN Pricing Delay", in: *The Jakarta Post* (22 October 2010), at: http://www.thejakartapost.com/news/2010/10/22/indonesia-geothermal-program-hung-pln-pricing-delay.html (16 March 2011).

[32] "Follow the Money: $400 Million Indonesian Commitment Has Players Scurrying", in: *Geothermal Digest* (29 March 2010), at: http://geothermaldigest.net/blog/2010/03/29/follow-the-money-400-million-indonesia-commitment-has-players-scurrying/ (16 March 2011).

[33] Padden, Brian, "World Bank Invests $400 Million in Indonesian Geothermal Energy", in: *Voice of America* (23 March 2010), at: http://www.voanews.com/english/news/World-Bank-Invests-400-Million-in-Indonesian-Geothermal-Energy–88906002.html (16 March 2011).

[34] ADB financed the plant as one of 12 subprojects under its Renewable Energy Development Sector (REDS) Project, aiming to increase the electricity output from Lahendong geothermal plant to 158 GWh (gigawatts hour) annually into PLN's Minhasa system of North Sulawesi; see World Bank, "ID-PCF-Indonesia Lahendong Geothermal Project" (25 August 2009), at: http://web.worldbank.org/external/projects/main?pagePK=64283627&piPK=64290415&theSitePK=40941&menuPK=228424&Projectid=P096677 (16 March 2011).

[35] JICA's contribution of ¥5,866 million (approximately US$70 million), beginning in March 2004, involved building a new plant with a 20 MW capacity that is due for completion in 2012; see JICA, "Major Projects: Lahendong Geothermal Power Plant Project", in: *JICA*, at: http://www.jica.go.jp/indonesia/english/activities/activity13.html (16 March 2011).

considered the substitution of existing and planned diesel plants in the West Nusa Tenggara, East Nusa Tenggara, Maluku and North Maluku provinces.[36,37]

5.3.4 Nuclear Energy

Finally, Indonesia is researching nuclear power, with three separate proposals currently under consideration.[38] Russian nuclear plant vendor, Rosatom, has proposed a small floating nuclear plant to supply Sulawesi.[39] Another project, supported by the International Atomic Energy Agency (IAEA), involves a nuclear-powered desalinisation plant on Madura that will generate electricity for the Java-Madura-Bali electricity grid.[40] The third plan is a revised proposal from the National Atomic Energy Agency (NAEA) for a major nuclear electricity generating station on the north coast of Central Java close to Mount Muria.[41] In their third attempt to receive government backing since the early 1980s, the agency has proposed building the first of four 1,000 MW pressurised water reactors. The World Bank and the Ministry of Finance opposed the first attempt while the second was aborted due to the Asian financial crisis of 1997.[42] Mitsubishi is expected to be one of the principal bidders should the Muria proposal go ahead.[43]

It is difficult to predict whether nuclear energy will materialise as a viable option for Indonesia in the near future, as government reports have often contradicted optimistic external prognostications. For example, in 2010, Energy and Mineral Resources Minister, Darwin Zahedy Saleh, when speaking about developing nuclear energy for electricity generation said, "[w]e will exploit other available energy sources first".[44] There are strong antinuclear groups that distrust

---

[36] "Pre-feasibility Study for Geothermal Power Development Projects in Scattered Islands of East Indonesia", in: *Engineering and Consulting Firms Association, Japan* (March 2008), at: http://www.ecfa.or.jp/japanese/act-pf_jka/H19/renkei/wjec_indonesia.pdf (16 March 2011).

[37] Alfian; Desy Nurhayati, "PLN Needs to Subsidize Geothermal Power", in: *The Jakarta Post* (5 May 2010), at: http://www.thejakartapost.com/news/2010/05/05/pln-needs-subsidize-geothermal-power.html (16 March 2011).

[38] "Indonesian Nuclear Power Proposals", in: *Nautilus Institute for Security and Sustainability* (7 February 2010), at: http://www.nautilus.org/initiatives/aus-indo/aust-ind-nuclear/ind-np/contemporary/#introduction (16 March 2011).

[39] Ibid.

[40] Ibid.

[41] Ibid.

[42] Ibid.

[43] "Mitsubishi Heavy Industries", in: *Nautilus Institute for Security and Sustainability* (19 January 2010), at: http://www.nautilus.org/initiatives/aus-indo/aust-ind-nuclear/ind-np/muria/mitsubishi-heavy-industries (16 March 2011).

[44] Nani Afrida, "Indonesia Will Not Develop Nuclear Power Anytime Soon: Minister", in: *The Jakarta Post* (2 May 2010), at: http://www.thejakartapost.com/news/2010/02/05/indonesia-will-not-develop-nuclear-power-anytime-soon-minister.html (16 March 2011).

the Indonesian government and doubt its ability to manage nuclear technology (Amir 2009, p. 353). This is especially so now that the politics and risk management of the nuclear option have become even more costly and complex in the aftermath of the 2011 Fukushima nuclear power plant disaster in Japan. In any case, even in the most optimistic fast-tracking scenario, nuclear power is not expected to make a significant contribution to Indonesia's energy security before 2030 (Hawksworth 2006, p. 6).

5.3.5 Indonesia's Position Vis-à-Vis the Trilemma

These findings emphasise that although coal and biofuels may seem attractive as options that ease Indonesia's energy security troubles, the increase in emissions from coal-fired power plants and the conversion of forests and peatlands for biofuel production will further contribute to climate change to which Indonesia is vulnerable for a variety of reasons (Fredriksson et al. 2007, p. 1824). These fuels are also likely to create environmental and social problems, especially through their contribution to climate change, impact on the living standards and social conditions of poor people and, in the case of biofuels, biodiversity loss due to the destruction of forests. While Indonesia is not in control of its climate change destiny, given the public good nature of the problem, it arguably has vested interests in mitigation initiatives and in setting an example that other developing countries might follow.

Indonesia has made pledges and commitments on environmental and energy policy changes that have placed the country in the spotlight as a pioneer of climate change mitigation through the promotion of renewable energy. The energy challenge for Indonesia, however, is a particularly complex one, as it has to not only reconcile energy security and climate change mitigation concerns but also deal with the legitimate claims of its citizens to not be plunged into energy and economic poverty as a result of national policies. Yet, many of the policies that aim to achieve energy security may threaten to remove ordinary Indonesians from their land or may exacerbate climate change and lead to increased flooding, landslides, water- or vector-borne diseases, windstorms, forest fires, drought and high tides/ storm surges (Ministry of Environment 2009, p. 12). Moreover, forest and peatland conversion to oil palm has resulted in other environmental and social impacts, which, in the short term, may cause higher commodity prices that could negatively affect access to food for poor people (Colbran and Eide 2008, pp. 4–10). Managing such a trilemma is difficult because, in Indonesia's case, it is difficult to find policies that avoid energy poverty—fuel subsidies cannot simply be swept away; investing in geothermal sources produces obvious gains for energy security and the climate, but does not do much for the millions of poor that are not on the grid and require investments into off-grid solutions; investing in coal meanwhile will do more to help supply electricity to those in urban settings than those in rural areas. If nothing were to be done to supply electricity to the poor, their chances of finding ways to increase their income would be further reduced.

The difficulty facing social, environmental and energy policymakers in Indonesia then is to determine how the country can develop its energy sector without causing additional negative climate, environmental and social impacts. Concerns regarding energy and social poverty are now so substantial and overt that they can no longer be ignored, either in social equity or political terms. A solution to this energy trilemma would be to develop energy that is reliable, affordable and sustainable. But, this is easy to say and indescribably difficult to achieve. At its heart, of course, is the sustainable development discourse and the implacable opposition of developing countries to sacrifice their future development and social and energy policy aspirations to an overweening goal of climate change mitigation.

Recognising the trilemma is one important step, and a second is to look for policy complementarities that can minimise adverse consequences across the prongs of the trilemma for Indonesia. Where this is not possible, policymakers will have to make some tough choices. For example, micro-hydropower projects that can provide off-grid solutions for those in rural areas are consistent with climate goals, but make very little contribution to meeting the scale of Indonesia's energy demands. The use of coal, at least in the short term, will contribute to Indonesia's energy security, but does not help it to reach its climate change goals and does nothing to relieve the energy poverty of those not on the grid. Working out the available policy complementarities requires, amongst other things, a fine-grain analysis of the country's diverse renewable energy forms, at least some of which could provide viable options for escaping the worst consequences of the trilemma. Examples of these options are discussed in the next section. Although such analysis will almost certainly not reveal a suite of policies that neatly resolve the trilemma, it will help to identify policies that target the trilemma in terms of the appropriate problems that need to be tackled. The degree to which a given policy is able to contribute to the management of the trilemma will become the focal point for evaluating its worth.

5.4 Sustainability Options and the Trilemma

As earlier established, Indonesia is rich in a variety of renewable energy resources and has options that offer benefits that fossil fuels do not as an energy source. First, renewable energy is a clean source of energy that gives off low emissions and thus helps to avoid several related costs, including associated health costs (e.g. power plant pollution impacts adversely on neighbouring communities), social costs (e.g. local communities are displaced by extreme weather events or the destruction of forests) and transportation costs (solar and micro-hydropower resources are local while imported coal is not). Interestingly, if similar costs (so-called 'external costs' or 'externalities') associated with power generation from fossil fuels were internalised, the production costs of oil or coal-fired power plants might be higher than that generated from geothermal, wind or solar sources.

Second, by definition, is the inexhaustibility of renewable energy. As discussed in Sect. 5.2, Indonesia is now facing rapidly diminishing oil reserves; other forms of fossil fuels too will be depleted in the future. This is not the case for the country's renewable resources, which, as at 2011, appear to be abundant. For instance, Indonesia's geothermal energy potential is spread along the islands of Sumatra, Java, Sulawesi and Maluku, which can be explored and developed to partially substitute coal as the main source of energy for power plants. However, only 4.2 % of this potential is currently being used.[45] When it comes to hydropower, Indonesia has the potential to generate 75,625 MW of power from the source, with 458 MW coming from small-scale hydropower projects (Abdullah 2005, p. 121; UNDP Indonesia 2009, p. 1). As was the case with geothermal energy, only about 6 % (4.3 GW) of Indonesia's total hydropower potential has been utilised (Frost and Sullivan 2011, p. 12). Efforts are underway to develop this resource at various levels. For example, a 2011 Ramon Magsaysay Award recipient, Tri Mumpuni, was recognised for helping her organisation build 60 community-run micro-hydropower plants, with a capacity to generate 5–250 KW of electricity and provide electricity to 500,000 people in rural Indonesia.[46] The organisation's success gives hope to millions of Indonesians, showing that even complex challenges, such as providing energy to rural Indonesia, can be overcome.

Solar energy radiation too holds out the promise of providing a cheap source of renewable energy and, more importantly, electricity, which might do much to mitigate the conditions of an estimated 70–80 million Indonesians who lack this energy service. Solar photovoltaic (PV) systems have great potential to supply electricity to rural households which are never likely to be reached by a national grid that, on efficiency grounds, must focus on readily accessible and heavily populated areas (Abdullah 2005, p. 123). Solar cookers also have an important role to play in reducing household air pollution.

In the case of biomass, Indonesia has the estimated capacity to provide 146.7 million tons of biomass annually (equivalent to 470 GJ per year), mainly from its significant agricultural and forest areas (Abdullah 2002, p. 1). Most of the biomass energy resources come from rice husk (150 GJ annually), rubber tree trunk (120 GJ annually), cocoa waste (7 GJ annually) and coconut palm waste (67 GJ annually) (Abdullah 2002, p. 3). According to Panaka (2005, p. 3), Indonesia's biomass potential is nearly 49.81 GW while its installed capacity is 445 MW. Biomass energy, offers a cleaner energy source, especially for cooking, in contrast to wood and kerosene cookers, which when used indoors account for high levels of ill health and breathing disorders. Wood stoves also provide only 5 % efficiency (Abdullah 2002, p. 3). Biomass can also be used for thermal processes in small-and

[45] Alfian, "Indonesia Geothermal Program Hung Up on PLN Pricing Delay", in: *The Jakarta Post* (22 October 2010), at: http://www.thejakartapost.com/news/2010/10/22/indonesia-geothermal-program-hung-pln-pricing-delay.html (16 March 2011).

[46] Ardiansyah, Fitrian, "Magsaysay and the Environment", in: *The Jakarta Post* (7 August 2011), at: http://www.thejakartapost.com/news/2011/08/07/magsaysay-and-environment.html (20 August 2011).

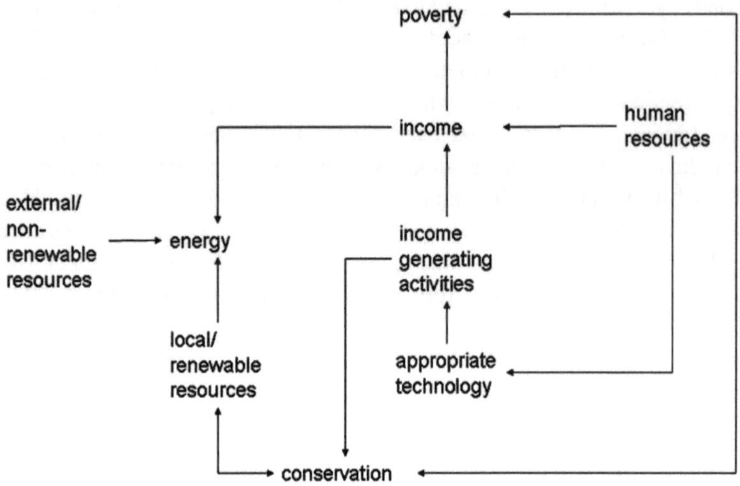

Fig. 5.2 Roles of agro-energy in promoting conservation and poverty alleviation. *Source* Ardiansyah and Suhud (2007, p. 2)

medium-scale industries, to generate electricity for equipments, such as rice driers, fish or cocoa driers and boilers, or as an energy source for other mechanical tools. It can also be converted to ethanol or biodiesel. Utilising biomass from rural agricultural by-products and waste can not only generate agriculture-based income but also provide increased energy services to rural households (see Fig. 5.2).

Proper utilisation of agro-energy could help prevent encroachment of forests and the uncontrolled collection of firewood and timber for income by large numbers of people by providing them with alternative energy and economic activities. It would also promote higher levels of ownership among local communities of forest areas by pegging forests as the provider of resources that people will likely want to safeguard. A similar logic can be applied to promote micro-hydropower and, indeed, also be used to persuade the government that the development of sustainable bioenergy can be scaled up through relatively low-scale investment. Tackling energy poverty through micro-projects will enhance the government's leverage when it comes to fuel subsidies reform; current subsidies, as noted earlier, generate long-term energy and fiscal insecurity.

To boost renewable energy development, the central government's general energy policy has advocated the diversification of energy sources and a conversion from coal and petroleum-based fuels to renewable energy sources to reduce emissions. For example, the Government of Indonesia has issued an energy policy (Law No. 30 of 2007), which stipulates that by 2025 the share of renewable energy should be 15 % of the national energy mix. To support this, the Ministry of Energy and Mineral Resources (ESDM) has launched a programme to provide electricity to isolated areas and small islands through, for example, solar PV (with a target of approximately 1,000 solar PV installations by 2008). Similarly, other projects on micro-hydropower and other small-scale renewables are also being developed gradually throughout Indonesia. Nonetheless, large-scale renewable energy development needs prioritisation in Indonesia in order to satisfy the significant increase in the country's energy demand, and this calls for the further development of geothermal energy, as it is one of the few renewable energy resources that can compete directly with coal for power generation. Geothermal energy can be used for baseload plant application, with a generation cost that is on par with most fossil fuels. This is reflected well in the second crash programme, outlined earlier, as part of which the government has targeted the private sector to finance and develop nearly 70 % of investments needed for geothermal power (Girianna 2009, p. 1).

Additional measures have also been put in place to allow Indonesia some time while it moves away from its fossil fuel dependence and toward using more renewable energy sources. These measures include energy management, energy saving and the wise use of electricity. The initiative is supported by an internal decree issued by the Indonesian President to government offices to undertake energy-saving steps. With the President's support and the increasing willingness of private sector actors to follow suit, Indonesia's measures to save energy and promote energy efficiency can be expected to contribute significantly to reducing its emissions in the interim.

In the specific context of large-scale bioenergy development with regard to palm oil, a 2007 WWF report suggests that the use of tropical fallow land, especially for planting oil palm is clearly more effective in terms of CO_2 savings than clearing of natural forests (Reinhardt et al. 2007, p. 23). Results are not as unequivocal when converting other plantations into oil palm, as it depends on the preceding crop and the substitution effects that flow from its replacement.

Strategic interventions could help to ensure positive GHG deliverance in palm oil development. However, such initiatives dictate that national and local governments adopt a cross-sectoral approach to bioenergy, fostering greater policy coherence among the agriculture, forestry and energy sectors in order to address the risk of conflict between competing land uses and to adopt the most appropriate forms of sustainable energy. This will have to be based on renewable energy supply actions following careful assessment of current and future demand and supply potentials.

A possible concrete and immediate intervention could be reviewing the permit allocation process for oil palm developments. This is an essential step that needs to be undertaken to ensure that these plantations no longer result in deforestation, peatland conversion and degradation. A review of the permits will require strong cross-sectoral coordination. For instance, on 20 May 2011, the Indonesian President signed a presidential instruction (No. 10 of 2011) regarding a two-year moratorium on new permits to clear primary forests and peatlands throughout Indonesia. This long-awaited moratorium, which was intended to reduce deforestation, may provide a venue for pushing not only the overall programme of Reducing Emissions from Deforestation and Forest Degradation Plus (REDD+) but also proper land use management and development in Indonesia.

Another way to develop oil palm without having to sacrifice forests is by increasing yields of smallholder plantations. In Indonesia, in terms of palm oil yields, smallholders tend to produce only an average of 2.3 tons per ha while private estates can produce up to 3.4 or even 8 tons per ha (Indonesian Forest Climate Alliance 2007, p. 32). Furthermore, the gap between yields on truly well-managed and high-yielding private estates and the national average for their group is much larger than it should be. Private producers have recorded 6.5–8.0 tons per ha average yields on individual plantations while the average for the group is little better than 4.1 tons per ha.[47] Poor smallholder yields are attributed to difficulties in obtaining good quality seedlings, incorrect plantation management and lack of sufficient capital to purchase fertilisers, pesticides and herbicides. If these productivity issues were addressed and improved, there would be no need for the palm oil sector to expand further, as all growth in demand will be met by the improving yields on existing plantations by 1.5–2 % annually (Unilever 2007, p. 1). The intervention will potentially reduce the need to convert forests and peatlands to oil palm. Income increases due to the improving productivity of smallholder plantations will also in turn provide some scope for improving energy services and markets.

In sum, the overall options for sustainability need much more detailed investigation, as no one actor in Indonesia has all the information or evidence needed to effectively manage the trilemma. The crippling effects of the trilemma can only be circumvented through a detailed evidence-based search for policy complementarities involving all parties, be it the government, the industry, civil society or other relevant actors. Nevertheless, a huge opportunity to decouple Indonesia's emissions from its energy development does exist.

5.5 A Transaction Cost Approach to the Trilemma

The authors have hitherto argued that Indonesia's energy policy should be conceived within a unifying framework for managing an energy trilemma rather than as a siloed policy process for improving energy security. In this section, the authors propose a general approach for implementing the policies that could contribute positively to managing the trilemma. For example, improving investment in geothermal resources is a plausible policy candidate for managing the trilemma. However, in a world where states have to compete for foreign direct investment and China outcompetes all, there remains the question of how Indonesia may implement a pro-geothermal policy. Our general suggestion here is that Indonesia should look for an approach that will reduce the transaction costs of

[47] "Indonesia: Palm Oil Production Growth to Continue", in: *USDA (United States Department of Agriculture)* (19 March 2009), at: http://www.pecad.fas.usda.gov/highlights/2009/03/Indonesia/ (16 March 2011).

implementation. This should not be confused as an argument for deregulation, but should rather be seen as a call for the removal of uncertainties around regulatory decisions already taken and their replacement by efficient executive motors of implementation.

Transaction costs are the costs of searching for, negotiating and enforcing contracts. For the present subject, our focus remains the search and information dimensions of deal-making in the renewable energy sector. Well and truly, before any formal contract is signed, developers in the renewable energy sector have to gather information about the best opportunities and partners available to them. There is an important broader context here. There are many renewable energy investment opportunities available globally to investors as both developed and developing countries, driven by concerns over energy security, begin to set national goals to increase the amount of energy they derive from renewable sources. Korea, for example, has announced, as part of its green growth plan, that it will increase its renewable energy supply from 2.7 % in 2009 to 3.78 % by 2013 and 6.08 % by 2020 (Presidential Commission On Green Growth 2009, p. 13). Similarly, India launched the Jawaharlal Nehru National Solar Mission in 2009 with an aim to provide 22,000 MW of solar power by 2022 (Global Energy Institute 2011, p. 11). There are many more examples that could be cited here. As the global demand for renewable energy initiatives continues, one plausible assumption is that investors will go to those countries where the transaction costs are lower.

Our interviews suggest that external multilateral investment agencies, such as the World Bank, are confronted by an opaque mix of political and multilevel regulatory authorities in Indonesia that generates uncertainty and entails transaction costs for investors. One source of these transaction costs is the division of power between central and local governments. With the current decentralised system, local governments have been given the rights and responsibility to issue concessions and operating licenses for renewable energy. However, most local governments have very limited capacity and understanding of the implications of various energy scenarios and there is no established procedure through which to pursue such initiatives.

Steps that have been taken to improve the quality of energy policy seem to have raised rather than lowered transaction costs. For instance, DNPI, which was established after the Bali Climate Change Conference in 2007, includes 16 cabinet members and involves six working groups of governmental officials to deal with issues of adaptation, mitigation, technology transfer, finance, forestry and post-Kyoto aims.,[48,49] While DNPI forms a new focal point for climate change strategy, it remains entirely unclear as to what its intended role is with regard to energy

[48] "National Action Plan: Addressing Climate Change", in: *State Ministry of Environment* (November 2007), at: http://www.uncsd2012.org/rio20/content/documents/Indonesia%20National%20Action%20Plan%20Addressing%20Climate%20Change.pdf (16 March 2011).

[49] Melisa, Eka, "National Action Plan and International Partnership: Indonesia's Response to Global Efforts to Improve the International Regime on Climate Change", in: *National Council on Climate Change* (2010), at: http://www.crawford.anu.edu.au/accpforum/pdf/ppp/7_Melisa.pdf (16 March 2011).

governance. Announcements that it will involve "building local capabilities", raise awareness of climate change threats and opportunities and effectively "coordinate Indonesia's response to climate change", are still typically preceded by statements, such as "building the institutional capability to ...", 3 years after DNPI's founding.[50]

Even in sectors where one might have expected transaction costs to have been streamlined by virtue of time and regulatory experience, there are problems. For example, Indonesia's oil and gas industry stalled in 2010, with a 25 % fall in state revenue from the sector expected. This was attributed to a failure to realise the sale of a large number of oil and gas blocks. The blame was allocated to erratic local government behaviour, failure of the upstream regulator and the bureaucracy's lack of technical capacity. One Jakarta-based development economist, for instance, described this by saying, "the state bureaucracy has a genius for producing more obstacles or disincentives".[51] This is also reflected in a 2010 survey of Asia's bureaucracies, which rated Indonesia the second worst.[52]

Faced by such transaction cost scenarios, one would expect that investors would react by going instead to China or perhaps Korea. Of course, Indonesia is not the only large developing country that generates high transaction costs for investors through the opaqueness of its political and regulatory system. India is another example. If Indonesia, for instance, were to move to find ways to reduce these transaction costs, then it might provide at least one reason for foreign investors in renewables to prefer Indonesia to India.

In the meantime, the development of renewable resources is moving very slowly. ESDM statistics confirm that, at present, renewable energy (hydropower, geothermal and biomass) accounts for only 3.4 % of total potential reserves.[53] Large-scale renewable energy development requires large-scale capital investment, and the Indonesian government is looking to the private sector for this investment. However, such investment is not coming in at the scale required at present and this should remain the case if our transaction cost hypothesis is correct. As one senior policymaker, bemoaning the lack of large-scale investment in renewables in Indonesia, said, "we need commitment to invest in a large way".

[50] "National Council on Climate Change: Low Carbon Development Strategies 03/06/2010", in: *Norway* (7 July 2010), at: http://www.norway.or.id/Embassy/development/Indonesia/environment/National-Council-on-Climate-Change-Low-carbon-development-strategies-03062010/ (16 March 2011).

[51] Lacey, Terry, "Bureaucracy Slows Indonesia Oil and Gas Development", in: *Asia Sentinel* (12 January 2010), at: http://www.asiasentinel.com/index.php?option=com_content&task=view&id=2236&Itemid=226(16 March 2011).

[52] "Indonesia's Bureaucracy among the Worst in Asia: Survey", in: *Jakarta Globe* (2 June 2010), at: http://www.thejakartaglobe.com/home/indonesias-bureaucracy-among-worst-in-asia-survey/378341 (16 March 2011).

[53] Ardiansyah, Fitrian, "Renewing Support for Renewable Energy", in: *The Jakarta Post* (2 February 2010), at: http://www.thejakartapost.com/news/2010/02/02/renewing-support-renewable-energy.html (16 March 2011).

Indonesia can, at least partly, improve its chances of securing this investment commitment if it focuses on governance strategies to reduce transaction costs being generated by layers of its bureaucracy. The use of national councils, such as DNPI and the Dewan Energy Nasional (DEN; also known as National Energy Council), may be useful inter-agency tools for deliberating policy and formulating plans. However, this appears, at least to outsiders, to have added another confusing layer to energy policy development in Indonesia. Understandably, investors would prefer to negotiate with a small stable set of bureaucratic decision-makers having executive powers of implementation instead of a large network of different players. This does not, however, mean that power should be limited to just one department, as extreme centralisation can also create problems of its own. Inter-agency processes involving energy regulators, such as BP Migas (the regulator for gas and oil), and other ministries, such as the Ministry of Finance, are crucially important to energy planning. As the authors have indicated, managing the trilemma will require the participation of many planners and experts from both within and outside the bureaucracy, and no one department should be allowed to direct energy policy—the trilemma is too complex and demanding in terms of information for that to be an efficient solution. Bureaucratic reform to improve planning processes, however, should not compromise the capacity of the bureaucracy to get things done once plans have been formulated, as this is generally a problem in Indonesia. As one interviewee commented, "BAPPENAS sets targets, but how are the targets to get funding? We have to seek support to get the money—either from our internal budget, from donors or from the private sector. And, as regards geothermal, we have to go to the Ministry of Energy to deal with technology and governance. And, the local government has licensing powers and do the transactions with private companies". In short, our suggestion is that Indonesia might look at what it can do to create a stable core of bureaucratic decision-makers who can fill the implementation gaps in Indonesia's energy policy, as processes of authoritative implementation are key to gaining the confidence of investors.

Most interviewees commented upon the rising importance of local governments in Indonesia's energy policy plans. Decision-making has been made immeasurably more complex by the devolution of power in 2003 to the hundreds of districts and municipalities in its 33 provinces, as part of the push for democratisation and regional autonomy following the collapse of Suharto's regime. This devolution has important implications for energy policy, for district governments have been given the right and responsibility to issue not only concessions but also operating licenses for renewable energy and energy efficiency. However, most districts and municipalities have limited capacity and understanding of the various energy scenarios and their implications, and there are no established routes or procedures to pursue such initiatives (IEA 2008, p. 33, 67). Moreover, opportunities for corruption have increased, and indeed corruption is prevalent in many such institutions, because local and regional institutions were unprepared for the transfer of power (Resosudarmo et al. 2000, p. 336).

The central government's macro plans for energy diversification call for close cooperation between district and provincial governments that will inevitably continue to think in local rather than national terms. Control over access to land and the various permit systems create opportunities for rent-seeking behaviour of all kinds. Officials at this level of government might also be risk averse when it comes to embracing opportunities for renewable energy projects, as it would be much easier to behave opportunistically and continue working with Indonesian coal companies that, in the words of one interviewee, are "printing money by selling to China". Indeed, the coal industry has had many years of experience in identifying relevant regional government mechanisms and officials through which business might be transacted, and of identifying circumstances and individuals where 'informal' payments may advance negotiations. Solving such problems may not be easy, but they will have to be solved if Indonesia is ever to drop the word 'potential' when it comes to describing its geothermal reserves of energy. Interestingly, all interviewees were acutely aware of the problems facing Indonesia at the local level of government when it came to implementing the strategic vision of its Energy Law of 2007.

Solving these problems will require patience on the part of the central government and negotiation with local governments using whatever financial levers it has at its disposal. One possibility is for the central government to create forums through which multilateral agencies, developers and banks meet with local government officials to explain the scale of funding and investment benefits available to local governments to help make geothermal, or other renewable energy options, a reality for Indonesia. The forum would need to be national in scope, so as to maximise its transparency, and operate under the leadership of the Indonesian President to give it convening power. The basic idea would be to use the forum to trigger a race for renewables, with those who move first at the local government level getting the biggest rewards. Creating such a forum would also help to lower the costs associated with searching for investors. Nationally publicising the success of a local government in, for example, a solar cooker initiative for poor people through the forum would be a good example of the kind of reputational tool that it could potentially be. A forum of 'carrots' could help to stimulate a race among local governments for renewable energy development, especially if driven by the stick of pricing reforms for fossil fuel energy.

5.6 Conclusion and the Way Forward

Indonesia's diminishing oil reserves and declining production capacity are putting pressure on the government to immediately come up with a plan to achieve energy security. We have presented that energy security is just one crucial prong of a policy trilemma facing Indonesia, the other two being climate change obligations and energy poverty. Conceiving the problem as managing a trilemma will encourage a more unified approach to its problem solving. The trilemma offers a basic guide to policy-making—its successful management requires a search for policy complementarities that will allow Indonesia to move forward on energy security, energy poverty and climate change. Managing the trilemma requires a fine-grained and evidence-based search and assessment of what is possible in the country's renewable energy sector. Policies that address only one prong of the trilemma will probably end up deepening the adverse consequences for the other prongs. For example, continuing with the use of subsidised oil will only cost the country's economy and place Indonesia in competition for a scarce resource, which is hardly a recipe for energy security. Similarly, increasing the use of coal for meeting short-term demand will only create long-term lock-in problems on GHG emissions in addition to promoting the adoption of cheap but inefficient coal technology. Coal, at best, can only serve as a partial response to the severe problems of energy poverty facing Indonesia.

While there is no way to easily resolve the trilemma, there are better ways to manage it. The key, we believe, lies in reforming the bureaucracy using the reduction of transaction costs around the implementation of energy policy as a broad criterion for changes that are essential. Increasing networked governance capacity around energy policy, for instance, via the creation of various councils, such as DNPI and DEN, that serve to enhance the capacity of the Indonesian government to create broader networks around these policies is desirable. What is not desirable is to create confusion and uncertainty around the implementation of energy policy decisions. Decisions once taken should not be allowed to drift into bureaucratic cycles of revisitation and renegotiation that smother investors in a smog of uncertainty. Such a vicious cycle can only favour established industries, such as coal and gas, that know the formal and informal paths to tread in order to achieve their desired outcomes. External investors interested in new initiatives in the renewable energy sector will without doubt end up as losers in this contest, and the upshot will be crash-through energy programmes borne of desperate short-term planning that end up deepening Indonesia's crises. While improving the management of the trilemma is definitely within Indonesia's reach, a necessary prerequisite for such improvement would be bureaucratic reform that targets its organisational practices around the implementation of energy policy.

Acknowledgments The authors would like to thank Muhammad Suhud and Iwan Wibisono for taking up the challenge of preparing and presenting the earliest version of this chapter in 2008 and Desak Putu Adhityani Putri for providing research assistance.

References

Abdullah K (2002) Biomass energy potentials and utilization in Indonesia. Laboratory of energy and agricultural electrification, Department of Agricultural Engineering, IPB and Indonesian Renewable Energy Society [IRES], Bogor

Abdullah K (2005) Renewable energy conversion and utilization in ASEAN countries. Energy (Amsterdam) 30(2–4):119–128

Agustina CDRD, del Granado JA, Bulman T, Fengler W, Ikhsan M (2008) Black hole or black gold? the impact of oil and gas prices on Indonesia's public finances. Policy research working paper 4718, World Bank, Jakarta

Amir S (2009) Challenging nuclear: antinuclear movements in postauthoritarian Indonesia. East Asian Sci Technol Soc Int J 3(2–3):343–366

Ardiansyah F, Suhud M (2007) Bioenergy for poverty alleviation. WWF Indonesia, Jakarta
Ariati R (2007) National policy on bioenergy. Paper for the Asian science and technology seminar, Jakarta, Indonesia, 8 March
BAPPENAS (National Development Planning Agency) (2009) Reducing carbon emissions from Indonesia's peat lands. BAPPENAS, Jakarta
Colbran N, Eide A (2008) Biofuel, the environment, and food security: a global problem explored through a case study of Indonesia. Sustain Dev Law Policy 9(1):4–1, 65–67
DNPI (Dewan Nasional Perubahan Iklim) Indonesia (2010) Indonesia's greenhouse gas abatement cost curve. National Council on Climate Change, Jakarta
EPRINC (Energy Policy Research Foundation Incorporated) (2007) Ethanol part II: is a home-grown fuel policy undermining U.S. energy security? EPRINC, Washington, DC
ESDM (Ministry of Energy and Mineral Resources) (2010) Handbook of energy and economic statistics of Indonesia. ESDM, Jakarta
Fredriksson GM, Danielsen LS, Swenson JE (2007) Impacts of el niño related drought and forest fires on sun bear fruit resources in lowland dipterocarp forest of East borneo. Biodiver Conserv 16(6):1823–1838
Frost & Sullivan (2011) Global energy mega trends and renewable energy outlook for Indonesia. Media briefing, Jakarta, Indonesia, 30 March
Girianna M (2009) Renewable energy and energy efficiency in Indonesia. Paper for the ADB workshop on climate change and energy, Bangkok, Thailand, 26–27 March
Global Energy Institute (2011) The rising sun—a point of view on the solar energy sector in India. KPMG, India
Hawksworth J (2006) The world in 2050: implications of global growth for carbon emissions and climate change policy. PriceWaterhouseCoopers, London
Hooijer A, Silvius M, Wösten H, Page S (2006) PEAT-CO_2: assessment of CO_2 emissions from drained peatlands in SE Asia, Delft Hydraulics Report No. Q3943. Delft Hydraulics, Delft
IEA (International Energy Agency) (2008) Energy policy review of Indonesia. OECD/IEA, Paris
Indonesian Forest Climate Alliance (2007) REDDI: reducing emissions from deforestation and forest degradation in Indonesia: REDD methodology and strategies summary for policy makers. Ministry of Forestry, Jakarta
Koalisi Energi (Koalisi Energi untuk Pembangunan Berkelanjutan) (2005) Tanggapan Koalisi Energi terhadap Blueprint Pengelolaan Energi Nasional (PEN) 2005–2025 dan Usulan Paradigma Baru Pengelolaan Energi Nasional (energy coalition's responses towards the national energy blueprint 2005–2025 and a new proposal on national energy management), Jakarta, Koalisi Energi, WWF Indonesia
Koh LP, Wilcove DS (2008) Is oil palm agriculture really destroying tropical biodiversity? Conserv Lett Hoboken 1(2):60–64
Leitmann J et al (2009) Investing in a more sustainable indonesia: country environmental analysis, 2009, CEA series East Asia and Pacific region. World Bank, Jakarta
Ministry of Environment (2009) Summary for policy makers: Indonesia second national communication under the United Nations framework convention on climate change (UNFCCC). Ministry of Environment, Jakarta
Ministry of Finance (2009) Ministry of finance green paper: economic and fiscal policy strategies for climate change mitigation in Indonesia. Ministry of Finance, Australia Indonesia Partnership, Jakarta
OECD (Organisation for Economic Co-operation and Development), IEA (International Energy Agency) (2010) Energy poverty: how to make modern energy access universal? OECD/IEA, Paris
Pallone S (2009) Indonesia's oil crisis: how Indonesia became a net oil importer. J Int Policy Solut 10(3):1–10
Panaka P, (2005) The role of biomass for the energy sustainable development in Indonesia. Paper for the 1st biomass Asia workshop, Hiroshima, Japan, Jan 19–21
Presidential Commission on Green Growth of the Republic of Korea (2009) Road to our future: green growth: national strategy and the five-year plan (2009–2013). Republic of Korea, Seoul

Reinhardt G, Rettenmaier N, Gärtner S, Pastowski A (2007) Rain forest for biodiesel? Ecological effects of using palm oil as a source of energy. WWF Germany, Frankfurt
Resosudarmo BP, Subiman NIL, Rahayu B (2000) The Indonesian marine resources: an overview of their problems and challenges. Indonesian Q 28(3):336–355
Resosudarmo BP, Alisjahbana A, Nurdianto DA (2010) Energy security in Indonesia, ANU working papers in trade and development No. 2010/08, Australian National University, Canberra
Sa'ad S (2009) An empirical analysis of petroleum demand for Indonesia: an application of the cointegration approach. Energy Policy (Amsterdam) 37(11):4391–4396
Sheil, D, Casson A, Meijaard E, van Noordwijk M, Gaskell J, Sunderland-Groves J, Wertz K, Kanninen M (2009) The impacts and opportunities of oil palm in Southeast Asia: what do we know and what do we need to know? CIFOR occasional paper No. 51, Center for International Forestry Research, Bogor
Teoh CH (2002) The palm oil industry in Malaysia: from seed to frying pan. WWF Switzerland, Zurich
Timilsina GR, Shrestha A (2009) Transport sector CO_2 emissions growth in Asia: underlying factors and policy options. Energy Policy Amsterdam 37(11):4523–4539
Tindale S (2009) How to meet the EU's 2020 renewables target, policy brief. Centre for European Reform, London
UNDP (United Nations Development Programme) Indonesia (2009) Project facts: rural development with renewable energy, climate change and sustainable development. UNDP Indonesia, Jakarta
Unilever (2007) Food industry action on palm oil sustainability: Unilever statement. Unilever UK, London
USAID (United States Agency for International Development) (2008) Indonesia energy assessment. USAID, Washington, DC
Wilcove DS, Koh LP (2010) Addressing the threats to biodiversity from oil-palm agriculture. Biodiver Conserv (Dordrecht) 19(4):999–1007
World Bank (2009) Energy and climate change, policy brief 51752. World Bank, Jakarta

Abbreviations

ADB	Asian Development Bank
ANU	Australian National University
APEC	Asia–Pacific Economic Cooperation
BAPPENAS	Badan Perencanaan Pembangunan Nasional (National Development Planning Agency)
CARR	Centre for the Analysis of Risk and Regulation
CIA	Central Intelligence Agency
CO_2	Carbon dioxide
DEN	Dewan Energy Nasional (National Energy Council)
DNPI	Dewan Nasional Perubahan Iklim Indonesia (National Council on Climate Change)
EIA	Energy Information Administration
ESDM	Kementerian Energi dan Sumber daya Mineral (Ministry of Energy and Mineral Resources, Republic of Indonesia)
EU	European Union
GDP	Gross domestic product
GHG	Greenhouse gas

GJ	Gigajoules
GW	Gigawatts
GWh	Gigawatts hour
ha	Hectares
HCVF	High conservation value forests
IAEA	International Atomic Energy Agency
IEA	International Energy Agency
IFCA	Indonesia Forest Climate Alliance
JICA	Japan International Cooperation Agency
KW	Kilowatts
LNG	Liquefied natural gas
LPG	Liquefied petroleum gas
mmboe	Millions of barrels of oil equivalent
$MtCO_2e$	Metric tonnes (tons) of carbon dioxide equivalents
mtoe	Million tonnes of oil equivalent
MW	Megawatts
NAEA	National Atomic Energy Agency
OECD	Organisation for Economic Co-operation and Development
OHS	Occupational Health and Safety
PERTAMINA	Perusahaan Tambang dan Minyak Negara (state-owned oil and gas company)
PLN	Perusahaan Listrik Negara (the state-owned electricity company)
PV	Photovoltaic
REDD+	Reducing Emissions from Deforestation and Forest Degradation Plus
REDS	Renewable Energy Development Sector
RSPO	Roundtable on Sustainable Palm Oil
tbd	Thousand barrels per day
tons	Metric tonnes
UK	United Kingdom
UN	United Nations
UNFCCC	United Nations Framework Convention on Climate Change
US	United States
USAID	United States Agency for International Development
USDA	United States Department of Agriculture
WWF	World Wide Fund for Nature

Author Biographies

Fitrian Ardiansyah (Indonesia/Australia): He is a climate and sustainability specialist; and former programme director for climate and energy, WWF Indonesia. He has over 15 years of experience in the field of environmental economics, natural resource management, integrated spatial and land use planning, sustainable commodities as well as climate change and energy. He is currently finalising his doctoral research at the Crawford School of Economics & Government, the Australian National University (ANU), Canberra, Australia. Where climate and energy is concerned, he was an expert for the Indonesia Forest Climate Alliance (IFCA), a member of the Indonesian official delegates to the United Nations Framework Convention on Climate Change (UNFCCC) and an advisory board member of the Asian Young Leaders Climate Forum. He was previously adjunct lecturer at the Post Graduate School of Diplomacy, Universitas Paramadina, and a member of the executive board of the Roundtable on Sustainable Palm Oil (RSPO). He recently received the Australian Leadership Award and Allison Sudradjat Award from the Government of Australia. His latest publications include: "Risk and Resilience in Cross-border Areas", in: Elliott, L.; Caballero-Anthony, M. (Eds): *Human Security and Climate Change in Southeast Asia* (New York: Routledge, forthcoming in February 2012). He also regularly writes for reputable publishers in Southeast Asia.

Neil Gunningham (Australia): He has degrees in law and criminology from the Sheffield University, UK. He is a barrister and solicitor (ACT) and holds a PhD from the Australian National University (ANU). He joined the Regulatory Institutions Network (RegNet) at the ANU in January 2002 and is currently co-director of the National Research Centre for Occupational Health and Safety (OHS) Regulation. He was previously foundation director of the Australian Centre for Environmental Law at ANU, visiting and senior Fulbright scholar at the Center for the Study of Law & Society, University of California, Berkeley, and visiting fellow at the Centre for the Analysis of Risk and Regulation (CARR), London School of Economics. His work on OHS regulation has focused on the mining industry and on the relationship between management systems-based approaches, trust and workplace culture. The insights generated apply to other industry sectors and resonate for other areas of regulation. One particular concern (as of the Asia–Pacific Economic Cooperation [APEC] Ministers Responsible for Mining) has been OHS in the broader Asia-Pacific region. Other research agendas concern climate change governance—examining how individual states and key actors within them, international institutions, and key non-state actors perceive these challenges as well as their negotiating possibilities and options. Publications include: *Mine Safety: Law Regulation Policy* (Sydney: Federation Press, 2007); (with Kagan, R.A.; Thornton, D.): *Shades of Green: Business, Regulation, and Environment* (Stanford: Stanford University Press, 2003); (with Sinclair, D.): *Leaders & Laggards: Next-Generation Environmental Regulation* (Sheffield: Greenleaf Publishing Limited, 2002); and (with Grabosky, P.): *Smart Regulation: Designing Environmental Regulation* (Oxford: Oxford University Press, 1998).

Peter Drahos (Australia): He holds degrees in law, politics and philosophy and is admitted as a barrister and solicitor. He is a professor at RegNet and holds a Chair in Intellectual Property at Queen Mary, University of London. He is a member of the Academy of Social Sciences in Australia. Prior to joining the Australian National University, he was an officer of the Australian Commonwealth Attorney-General's Department, where he drafted Commonwealth legislation. He has published widely in law and social science journals on a variety of topics, including contract, legal philosophy, telecommunications, intellectual property, trade negotiations and international business regulation. He has served as a consultant to the government, international organisations and international non-governmental organisations. Publications include: *A Philosophy of Intellectual Property*, Applied Legal Philosophy Series (Aldershot-Brookfield, US: Dartmouth Publishing Group, 1996); (with Braithwaite, J.): *Global Business Regulation* (Cambridge: Cambridge University Press, 2000); (with Braithwaite, J.): *Information Feudalism: Who Controls the Knowledge Economy?* (London: Earthscan, 2002); (co-edited with Mayne, R.): *Global Intellectual Property Rights: Knowledge, Access and Development* (Hampshire: Palgrave Macmillan, 2002); and *The Global Governance of Knowledge: Patent Offices and Their Clients* (Cambridge: Cambridge University Press, 2010).

planning, sustainable commodities as well as climate change and energy. He is currently completing his doctoral research at the Crawford School of Economics & Government, the Australian National University (ANU), Canberra, Australia. Where climate and energy is concerned, he was an expert for the Indonesia Forest Climate Alliance (IFCA), a member of the Indonesian delegation to the United Nations Framework Convention on Climate Change (UNFCCC) and an advisory board member of the Asian Young Leaders Climate Forum. He was previously adjunct lecturer at the Post Graduate School of Diplomacy, Universitas Paramadina, and a member of the executive board of the Roundtable on Sustainable Palm Oil (RSPO). He recently received the Australian Leadership Award and Allison Sudradjat Award from the Government of Australia. His latest publication includes "[illegible] and Resilience in Cross-Border Areas" in Elliott, L., Caballero-Anthony, M. (eds) *Human Security and Climate Change in Southeast Asia* (2012). [illegible] writes for reputable publications [illegible]

[illegible]

About the RSIS Centre for Non-Traditional Security (NTS) Studies

The **RSIS Centre for Non-Traditional Security (NTS) Studies** conducts research and produces policy-relevant analyses aimed at furthering awareness and building capacity to address NTS issues and challenges in the Asia-Pacific region and beyond.

To fulfil this mission, the Centre aims to:

- Advance the understanding of NTS issues and challenges in the Asia-Pacific by highlighting gaps in knowledge and policy, and identifying best practices among state and non-state actors in responding to these challenges.
- Provide a platform for scholars and policymakers within and outside Asia to discuss and analyse NTS issues in the region.
- Network with institutions and organisations worldwide to exchange information, insights and experiences in the area of NTS.
- Engage policymakers on the importance of NTS in guiding political responses to NTS emergencies and develop strategies to mitigate the risks to state and human security.
- Contribute to building the institutional capacity of governments, and regional and international organisations to respond to NTS challenges.

Our Research

The key programmes at the **RSIS Centre for NTS Studies** include:

(1) Internal and Cross-Border Conflict

- Dynamics of Internal Conflicts
- Multi-level and Multilateral Approaches to Internal Conflict
- Responsibility to Protect (RtoP) in Asia
- Peacebuilding

M. Caballero-Anthony et al. (eds.), *Energy and Non-Traditional Security (NTS) in Asia*, SpringerBriefs in Environment Security, Development and Peace 1,
DOI: 10.1007/978-3-642-29706-9, © The Author(s) 2012

(2) Climate Change, Environmental Security and Natural Disasters

- Mitigation and Adaptation Policy Studies
- The Politics and Diplomacy of Climate Change

(3) Energy and Human Security

- Security and Safety of Energy Infrastructure
- Stability of Energy Markets
- Energy Sustainability
- Nuclear Energy and Security

(4) Food Security

- Regional Cooperation
- Food Security Indicators
- Food Production and Human Security

(5) Health and Human Security

- Health and Human Security
- Global Health Governance
- Pandemic Preparedness and Global Response Networks

Our Output

Policy Relevant Publications

The **RSIS Centre for NTS Studies** produces a range of output such as research reports, books, monographs, policy briefs and conference proceedings.

Training

Based in RSIS, which has an excellent record of post-graduate teaching, an international faculty, and an extensive network of policy institutes worldwide, the Centre is well-placed to develop robust research capabilities, conduct training courses and facilitate advanced education on NTS. These are aimed at, but not limited to, academics, analysts, policymakers and non-governmental organisations (NGOs).

Networking and Outreach

The Centre serves as a networking hub for researchers, policy analysts, policymakers, NGOs and media from across Asia and farther afield interested in NTS issues and challenges.

The **RSIS Centre for NTS Studies** is also the Secretariat of the Consortium of Non-Traditional Security Studies in Asia (NTS-Asia), which brings together 20 research institutes and think tanks from across Asia, and strives to develop the process of networking, consolidate existing research on NTS-related issues, and mainstream NTS studies in Asia.

More information on our Centre is available at http://www.rsis.edu.sg/nts

Research in the RSIS Centre for NTS Studies received a boost when the Centre was selected as one of three core institutions to lead the MacArthur Asia Security Initiative in 2009.

The Asia Security Initiative was launched by the John D. and Catherine T. MacArthur Foundation in January 2009, through which approximately US$68 million in grants will be made to policy research institutions over seven years to help raise the effectiveness of international cooperation in preventing conflict and promoting peace and security in Asia.

MIX
Papier aus verantwortungsvollen Quellen
Paper from responsible sources
FSC® C105338

If you have any concerns about our products,
you can contact us on
ProductSafety@springernature.com

In case Publisher is established outside the EU,
the EU authorized representative is:
Springer Nature Customer Service Center GmbH
Europaplatz 3, 69115 Heidelberg, Germany

Printed by Libri Plureos GmbH
in Hamburg, Germany